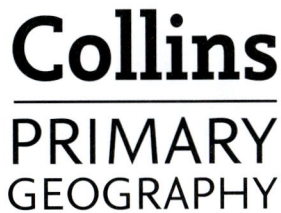

Issues
Pupil Book 6

Category	Topic	Page
Planet Earth	**Restless Earth**	
	Earthquakes and volcanoes	2
	Creating landscapes	4
	Rocks and soils in the UK	6
Water	**Drinking water**	
	Water, water everywhere	8
	Water supplies	10
	Conserving water	12
Weather	**Climate change**	
	Global warming	14
	Unusual weather	16
	Responding to climate change	18
Settlements	**Planning issues**	
	Reasons for development	20
	Old sites, new uses	22
	Planning game	24
Work and Travel	**Transport**	
	Travelling further, travelling faster	26
	Transport problems	28
	Hidden costs	30
Environment	**Conservation**	
	Threatened wildlife	32
	Antarctica	34
	Conservation projects	36
Places	**England**	38
	Europe	44
	South America	50
	Asia	56
	Glossary	62
	Index	63

Stephen Scoffham | Colin Bridge

Unit 1 Restless Earth

Lesson 1: Earthquakes and volcanoes

> What do we know about the Earth's crust?

The ground beneath our feet seems firm and solid, yet every so often earthquakes and volcanoes make it shake and crack. Earthquakes and volcanoes happen suddenly; other Earth movements happen gradually.

Sometimes fossil seashells are found in the rocks on high mountains. This proves to scientists that these rocks were once on the seabed.

▼ Earthquakes are measured by a seismograph. This graph shows how much the Earth moved during an earthquake on a Pacific island.

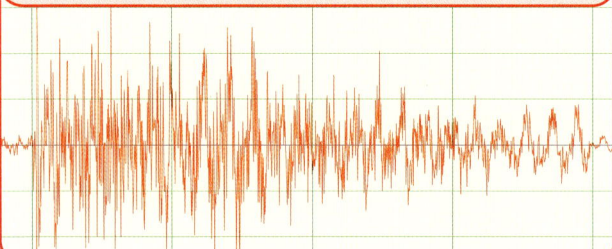

Discussion
- What clues show that some mountains are made of rocks that were once under the sea?
- What are the three sections that make up the Earth?
- Why might volcanoes be found in lines or groups?

▼ These layers of rock at Lulworth Cove in Dorset, UK, have been twisted and bent by the Earth's movements.

Key words
core
crust
earthquake
fossil
mantle
seismograph
volcano

Unit 1 Restless Earth

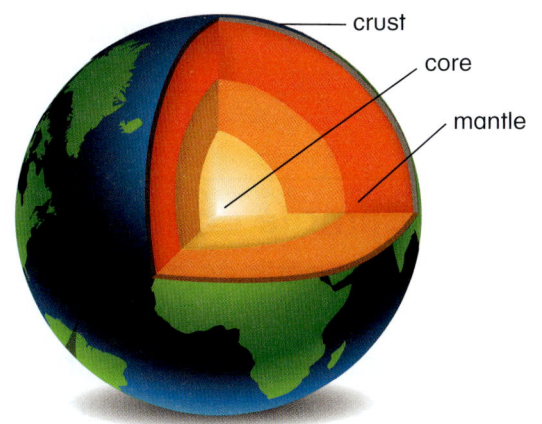

Data bank
- Between 50 and 70 volcanoes erupt each year – about one a week.
- Three-quarters of the world's volcanoes are in the 'ring of fire' around the Pacific Ocean.
- Nearly three-quarters of the energy used in Iceland comes from hot rocks under the ground.

The Earth is made up of three different sections. The surface, or crust, is between 6 and 40 kilometres thick. It consists of solid rock. Beneath the crust is a section called the mantle. Parts of the mantle are so hot that the rocks have melted and flow like a sticky liquid. The core of the Earth is an even hotter ball of iron and nickel.

▲ Earthquakes happen when two parts of the crust move apart or grind together. This picture shows a road cracked by an earthquake.

Mapwork
Working from an atlas or online, name six famous volcanoes. Say when they erupted, the country where it happened and the latitude and longitude.

▲ When a volcano erupts, it brings hot gases and rocks to the surface. If the lava goes on flowing, high mountains build up. This photograph shows a volcano in Iceland.

Investigation
Find out what happens when a volcano erupts and draw a diagram to illustrate the information.

Unit 1 Restless Earth

Lesson 2: Creating landscapes

What forces shape the land?

Key words
erosion
glacier
landscape
limestone

Although most rocks are very hard, they can still be cut and made into shapes. Sculptors carve blocks of stone to make statues. In a similar way, natural forces shape landscapes around the world. There are five main forces.

- **Frost** Rain runs into cracks in rocks. In cold weather the water freezes, breaking the rocks apart.
- **Rivers** Rivers carry particles of rock downstream creating valleys, hills and waterfalls.
- **Waves** Around the coast, waves undermine cliffs and wear away headlands.
- **Wind** Strong winds pick up particles of rock and blast them against cliffs and mountains.
- **Ice** In cold places, glaciers scrape away the rocks beneath them.

These processes all happen very slowly but over millions of years, even mountain ranges can be worn down to sea level. This is called erosion.

Climate change
- Coastlines are more unstable as climate change brings extreme weather and rising sea levels.
- At Holderness in the UK, two metres of coast is worn away every year.
- Rising sea levels could flood island nations such as Tuvalu and the Maldives by 2050.

◀ Cliff erosion threatens these houses.

Discussion
- What five forces shape the land?
- Where have you seen erosion wearing away surfaces?
- Will there be more erosion if climate change makes the weather wetter, drier or hotter?

Investigation
Find out how erosion is affecting your school. Take some photographs or make drawings of wear and tear resulting from natural processes and write captions explaining what has caused it.

Unit 1 **Restless Earth**

◀ In the desert, wind and water wear away the land. In Monument Valley, US, pillars of hard rock have been left behind.

▶ Glaciers scour out deep valleys, creating knife-edge ridges and jagged peaks such as the Matterhorn in Switzerland.

▼ The Niagara Falls on the St Lawrence River are moving upstream about one metre a year as the water wears away the rock.

▼ These caves in Croatia were created by water dissolving limestone rocks underground.

Unit 1 | Restless Earth

Lesson 3: Rocks and soils in the UK

How has the landscape of the UK formed?

Key words
clay
coal
coral reef
flint
granite
limestone
swamp

The landscape that we see around us today has been shaped over millions of years. In some places the land has been pushed up into mountains. In others it has been worn away. These processes are still going on.

Volcanoes

"I am a mountain guide in Eryri."

▼ Castell y Gwynt in Eryri (Snowdonia).

500 million years ago most of the UK was covered by sea. Underwater volcanoes erupted forming the mountains of Eryri (Snowdonia), Wales.

Deserts

"I am a farmer in Herefordshire."

▼ Fields of red soil, Herefordshire.

400 million years ago the land was pushed up out of the sea. Rivers deposited sand from the desert, making fertile soil.

Swamps

"I work on an oil rig in the North Sea."

▶ A North Sea oil rig.

300 million years ago places in and around the UK were covered by swamps. The trees and plants slowly rotted, creating coal, oil and gas.

Coral seas

"I am a house builder in the Cotswolds."

▼ A Cotswold village, Oxfordshire.

200 million years ago the seas covered the land again. Over time sea creatures and coral reefs turned into limestone which is now used in buildings.

Unit 1 Restless Earth

Rocks in the street

Rocks are valuable building materials. If you look around, you can see how people have used them in your local environment. You will probably be able to find rocks and stones which have been brought from other areas.

These are some of the clues to look for.

Rock	Way it is used
Limestone	Cut to make building blocks
Flint	Cemented together to make walls
Slate	Split into thin sheets for roof tiles
Clay	Baked in kilns to make bricks and tiles
Granite	Broken into chips to make road surfacing

brick chimney
clay tiles
flint walls
road made from tar and chippings

Mapwork
Devise a trail in or around your school. Put in stopping places where different types of rock can be found, including roads, pavements and flower beds.

Investigation
Build up a rock collection with labels on a display table. Ask everyone to contribute. Add things which have been made from rocks on a second table.

Climate change
What have people discovered about the climate over millions of years by studying rocks and soils?

Summary
In this unit you have learnt:
- about the way the Earth's crust moves
- about the processes which shape the landscape
- how rocks affect the character of places in the UK.

Unit 2 Drinking water

Key words
borehole
pumping station
reservoir
resource
waterworks
well

Lesson 1: Water, water everywhere

Is there enough water in the world?

There are huge quantities of water in the world. However, most of it cannot be used for drinking as it is salty seawater. Our main sources of fresh water are rivers, lakes and underground rocks.

Fresh water is essential to our lives as well as for factories and farms. Around the world, the demand for water is rising as the population increases. Also people are using more water as they buy more machines. This means that water is becoming a scarce resource. The problem is worse in crowded areas which have little rain.

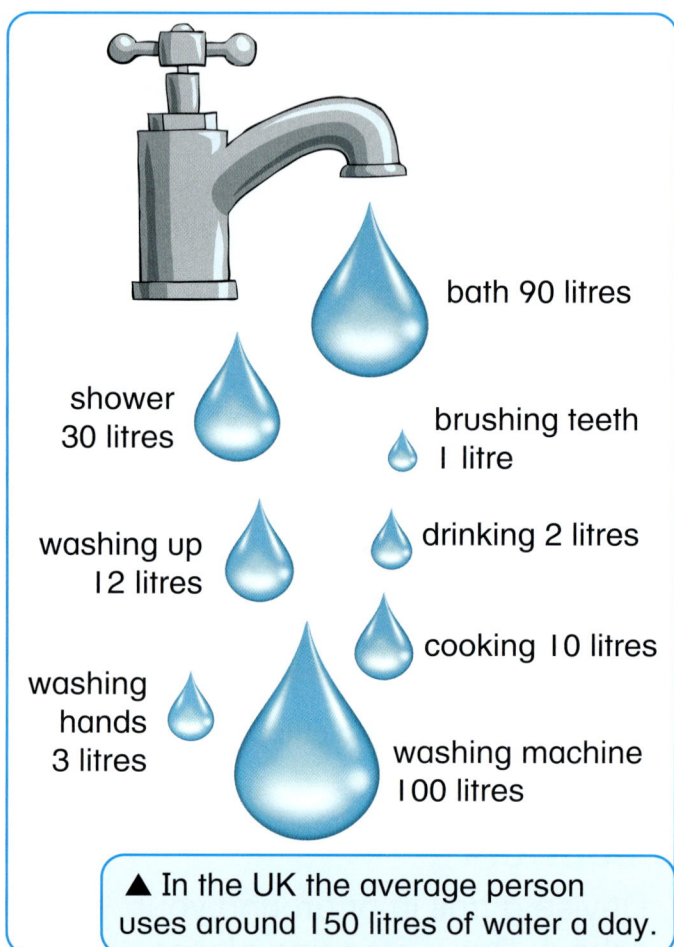

bath 90 litres
shower 30 litres
brushing teeth 1 litre
washing up 12 litres
drinking 2 litres
cooking 10 litres
washing hands 3 litres
washing machine 100 litres

▲ In the UK the average person uses around 150 litres of water a day.

Data bank
- It takes 200 million litres of water a second to grow all the world's food.
- Seventy per cent of fresh water is in Antarctica, frozen as ice.
- The water that flows down the River Thames in London, UK, may have been drunk by as many as eight people before it reaches the sea.

Data bank
- A dripping tap can waste 700 litres or more a month.
- Water meters have been installed in many homes worldwide to help save water.
- Around the world more people have a mobile phone than a flushing toilet.

Unit 2 Drinking water

In the UK, there is more rain than people need. The wettest areas are the mountains of Scotland and Wales where few people live. Some of the water is stored in reservoirs and pumped to other parts of the country which are drier and more crowded.

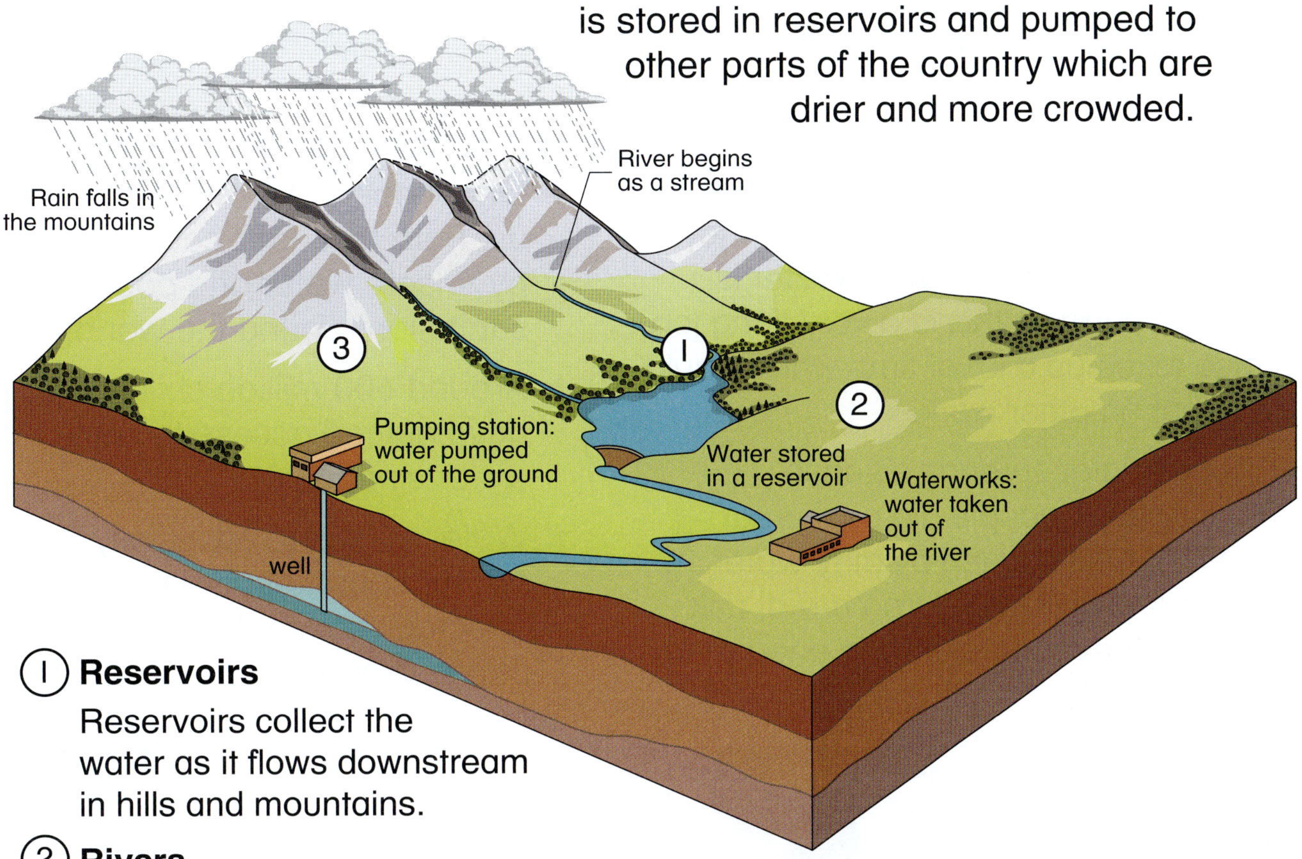

Rain falls in the mountains

River begins as a stream

③ Pumping station: water pumped out of the ground

well

① Water stored in a reservoir

② Waterworks: water taken out of the river

① **Reservoirs**
Reservoirs collect the water as it flows downstream in hills and mountains.

② **Rivers**
Sometimes water is taken straight out of rivers. It is cleaned at a waterworks and then pumped to our homes.

③ **Wells**
Wells or boreholes are built so we can reach water in rocks deep underground. The water is pumped to the surface.

▼ Kielder reservoir in Northumberland is the largest artificial reservoir in northern Europe.

Mapwork
Working from a map, draw a sketch map to show the region drained by your nearest river.

Investigation
Using data from the tap diagram, create a bar chart of water use.

Unit 2 Drinking water

Lesson 2: Water supplies

Why is clean water so important?

Around the world some people do not have water piped to their homes. Instead they have to collect what they need from a river or a well. Water is heavy to carry so this is a very tiring job. It also takes up a lot of time which could be spent on other work.

Polluted water spreads disease. Thousands of people die each year from illnesses caused by dirty water. Babies and young children are at risk because they are not strong enough to resist germs.

Key words
- dam
- pump
- well

Data bank
- One thousand children a day die from diseases caused by unsafe water and poor sanitation around the world.
- Around 600 million children have gained access to safe water since 2000.
- Water supplies, especially in poorer countries, are being affected as climate change brings droughts and floods.

▼ It takes a lot of time and energy to collect well water.

Discussion
- Where do some people have to go to get their water?
- Why is clean water so important?
- If you had to fetch water from a river or well, what three things would you use it for?

NORTH AMERICA EUROPE ASIA AFRICA SOUTH AMERICA OCEANIA

▶ Access to safe water is still an issue in parts of the world such as regions of Africa, Asia and North America.

Key
Places where many people do not have clean water

Unit 2 Drinking water

Improving water supplies

World leaders agree that everyone should have clean water.

> **Mapwork**
> Choose two photographs below. Draw diagrams to show where the water comes from and how it reaches the people who drink it.

> **Investigation**
> Using an atlas and the map on page 10, name six countries where many people do not have clean drinking water.

▲ **Piped water in China**
Pipes carry water to fields and villages in China.

▶ **Water tanks in India**
Tanks store rainwater from the roofs of houses. The water is used for washing and watering plants in the kitchen garden.

▼ **Dams in Bolivia**
Dams in the hills store water in the rainy season. These provide water for people when it is dry.

▲ **Wells in Kenya**
Pumps bring pure water to surface from rocks under the ground. The pump sucks it up to the surface.

Unit 2 Drinking water

Lesson 3: Conserving water

Key words
canal
crops
irrigation
monsoon
water supply

Are we using water wisely?

The River Indus rises in the region of Tibet in China, high up in the Himalayas. It flows over 3000 kilometres through mountains and deserts before it reaches the Arabian Sea. People, plants and animals depend on water from the Indus. In the past, there were great cities in the Indus valley. Today, the Indus supplies water for most of the crops in Pakistan.

There are three large dams and more than a hundred smaller dams along the Indus. These generate electricity and store water to irrigate the fields. They are also valuable for fishing. Today, climate change is slowly causing the Indus to become drier. It also suffers from plastic and chemical pollution.

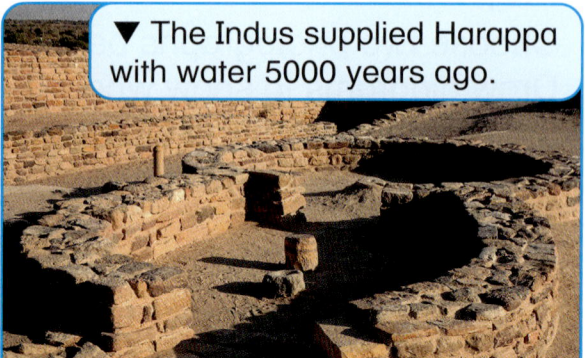
▼ The Indus supplied Harappa with water 5000 years ago.

▼ Canals distribute water to the fields.

▼ The Tarbela Dam makes electricity and stores water from monsoon rains.

Discussion
- Why is the River Indus important?
- What problems are affecting the River Indus today?

▶ A satellite image of the River Indus.

Mapwork
Make your own sketch map of the River Indus. Remember to add a key and a north point.

Unit 2 Drinking water

A water survey Children in two different schools decided to investigate if they were saving water.

School A   **School B**

Taps turn themselves off automatically. ①

Taps keep running if they are left on. ②

Toilets have small cisterns. ③

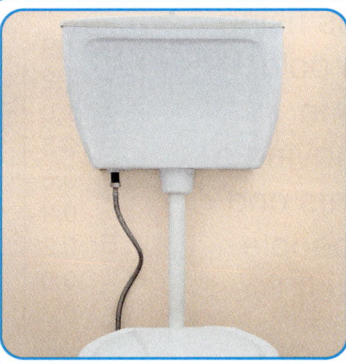
Toilets have large cisterns. ④

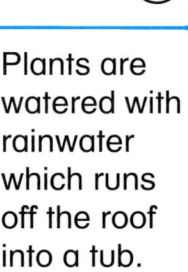
Plants are watered with rainwater which runs off the roof into a tub. ⑤

Rainwater runs straight down the drains and is not saved for other uses. ⑥

Investigation
Devise a survey to discover if people use water wisely in your school.

Summary
In this unit you have learnt:
- where drinking water comes from
- that polluted water causes illness
- how people can save water.

Unit 3 Climate change

Lesson 1: Global warming

> How is the world's climate changing?

Key words
climate
extreme weather
fossil fuels
global warming
polar regions
temperature
greenhouse effect

The climate is changing all over the world. Since around 1850, with increasing industrialisation, global temperatures have risen by over 1 °C. The world is hotter now than it has been for thousands of years and extreme weather events such as floods, droughts and storms are becoming more common.

Some places are warming more quickly than others. The temperature is rising particularly fast in polar regions. Changes in temperature are having a big impact on plants and creatures. Many people are also finding their lives are disrupted.

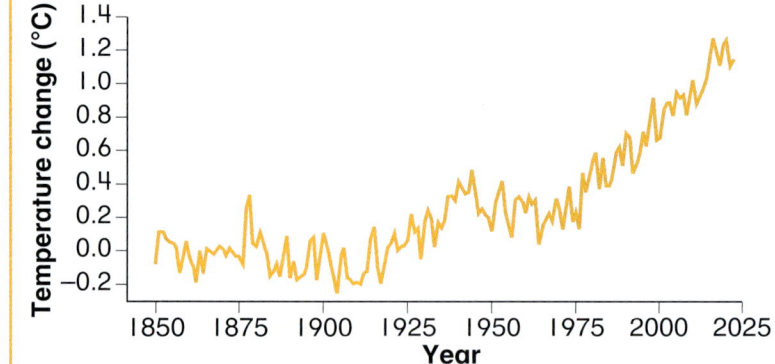

▼ Recent years have been the hottest ever recorded.

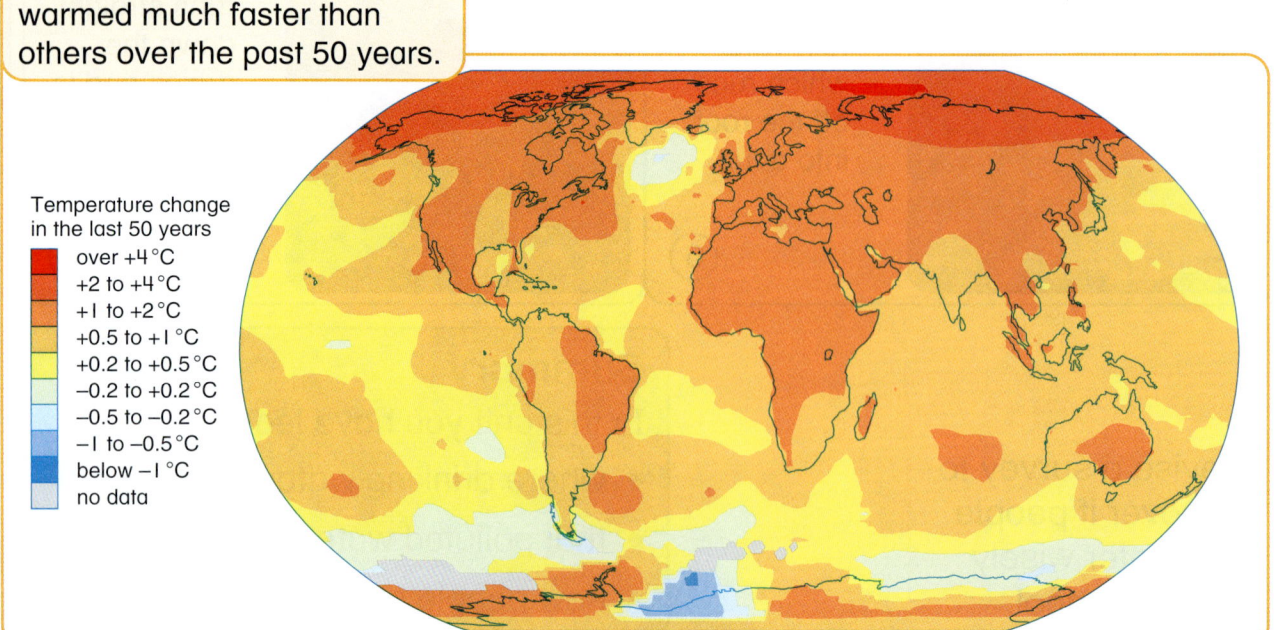

▼ Some places have warmed much faster than others over the past 50 years.

Temperature change in the last 50 years
- over +4 °C
- +2 to +4 °C
- +1 to +2 °C
- +0.5 to +1 °C
- +0.2 to +0.5 °C
- −0.2 to +0.2 °C
- −0.5 to −0.2 °C
- −1 to −0.5 °C
- below −1 °C
- no data

Unit 3   Climate change

The greenhouse effect

Scientists now know that global warming is caused by human activity. When we burn fossil fuels such as coal, oil and gas to make energy to run factories or drive cars, for example, it puts carbon dioxide gas into the air. This gas traps the heat from the Earth and stops it bouncing back into space. Carbon dioxide is one of several gasses that cause this to happen.

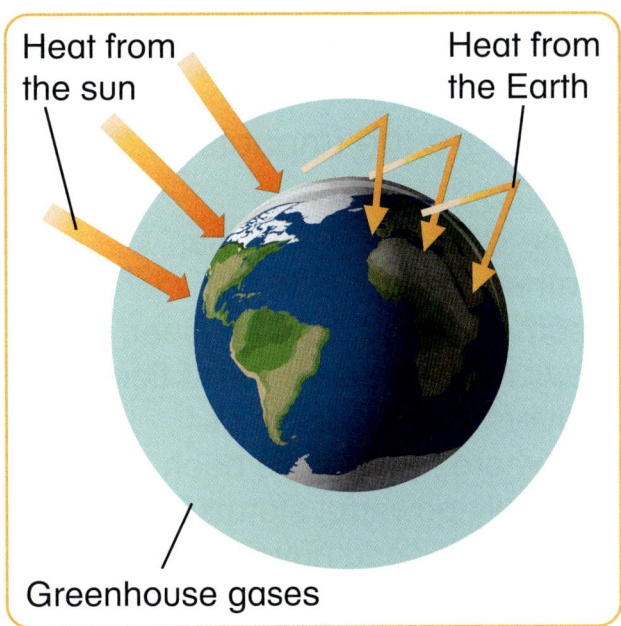

Heat from the sun

Heat from the Earth

Greenhouse gases

Climate change

If temperatures go on rising, it could be very serious. Governments around the world are working hard to stop this happening. They have passed laws to try to limit warming to 1.5 °C.

Impact of rising temperatures

4 degrees	Many coasts flooded
	Sea levels rise as Antarctic ice melts
3 degrees	Amazon begins to turn to desert
	Millions of people made homeless
2 degrees	Collapse of the Greenland ice sheet
	Lots of plants and animals endangered
1 degree	Arctic sea ice melts in summer
	More heat waves and forest fires

Mapwork

Look at the map on page 14. Why do you think land areas are warming faster than seas and oceans?

Discussion

- Why would it matter if the Greenland ice sheet collapsed?
- What do people gain from burning polluting fossil fuels?

15

Unit 3 Climate change

Lesson 2: Unusual weather

What are the impacts of global warming?

Key words
drought
flood
forest fire
hurricane

As the climate gets warmer, weather patterns are being disrupted. In some places there has been very heavy rain that has caused floods as rivers burst their banks. Elsewhere, there have been unusually powerful storms. Some parts of the world have suffered long periods of drought.

People in poorer countries suffer most from global warming. They do not have the money or materials to protect themselves properly. Plants and creatures are also threatened. They often cannot adapt fast enough to changing conditions. As the balance of nature is disrupted, many species are now at risk.

Climate change
- There are 38 small island states at particular risk from flooding and sea level rise.
- More than half of refugees in 2022 left their homes due to climate change.
- Climate change has happened before. About 65 million years ago, it wiped out the dinosaurs.

Mapwork
Look at a map of the world. List ten small island states and the oceans and seas where they are found.

Investigation
Research one of these animals and prepare a presentation about where it is found and why it is at risk.

Tiger

Giant panda

Polar bear

Albatross

Cheetah

Green sea turtle

Asian elephant

Unit 3 Climate change

▼ Forest fires
In Australia very hot, dry weather can cause serious bushfires.

▶ Floods
Heavy rain brought floods to this village in England.

▶ Storms
Hurricane Beryl hit the Caribbean in July 2024. Hurricanes are getting stronger as the sea gets warmer.

▼ Drought
A dry reservoir in Wales.

Discussion
Look at each photograph. Discuss the impact of each event on people, plants and creatures.

Unit 3 Climate change

Lesson 3: Responding to climate change

What can we do about climate change?

Key words
carbon emissions
carbon footprint
net zero
regenerative farming

We can reduce global warming if we burn less coal, oil and gas. Another way is to limit climate-warming gases from housing, transport and farming.

Net zero
Many countries have targets to reach 'net zero' by 2050. This means that any activity that puts carbon dioxide into the atmosphere has to be balanced by taking the same amount of carbon dioxide out. Emissions come from homes, factories, farming and transport. Trees absorb carbon dioxide so help to cancel them out.

Regenerative farming
Many modern farms use a lot of chemicals and energy that create pollution. Some farmers are trying hard to care for the soil and work in harmony with the seasons and nature. This reduces the need for chemicals, helps to keep animals healthy and cuts emissions. This is called regenerative farming.

NET ZERO
- These icons show some of the things we need to do to reach net zero. Can you work out what each icon represents?

18

Unit 3 Climate change

Net zero exhibition

A museum in Manchester, UK, asked children about the things that would no longer be used in 2050 if we succeed in reaching net zero. These are some of the things they put in their exhibition:

Baths	Gas boilers
Bottled water	Insecticides
Chainsaws	Petrol engines
Coal	Plastic toys
Disposable nappies	Throwaway clothes
	Tumble dryers

Carbon footprints

The amount of carbon dioxide that our actions create is called our carbon footprint. These are some of the ways to reduce your impact:

- walk or cycle if you can
- take care not to waste water
- eat less meat and dairy
- recycle things you do not want
- avoid buying things wrapped in plastic
- save energy by turning off lights
- grow plants to eat
- plant flowers for bees and insects
- don't buy things you don't need.

Investigation

What three things would you choose for your own net zero exhibition?

▶ Children helped create a place for wildlife for everyone to enjoy in this community garden.

Discussion

- What evidence is there in your area that people are trying to reduce carbon emissions? For example, solar panels, electric cars, wind farms.
- Five principles which guide climate change action are: rethink, reduce, reuse, repair, recycle. How could you do these things in your life?
- If you could only do one thing to reduce your carbon footprint what would it be?

Summary

In this unit you have learnt:

- how the world is getting warmer
- how global warming is impacting people, plants and creatures
- some of the things you can do in response to climate change.

Unit 4 Planning issues

Lesson 1: Reasons for development

Why are there conflicts over land use?

Key words
industry
leisure
sustainability
transport
warehouse
greenhouse

People use the land in many ways. Some areas are farmed and produce crops. In other places there are big cities where millions of people live. Sometimes several people have different plans for the same piece of land. A decision then has to be made about which scheme is best.

Investigation
Look at some old maps of your school and local area. Make a list of places which have been used in different ways in the past.

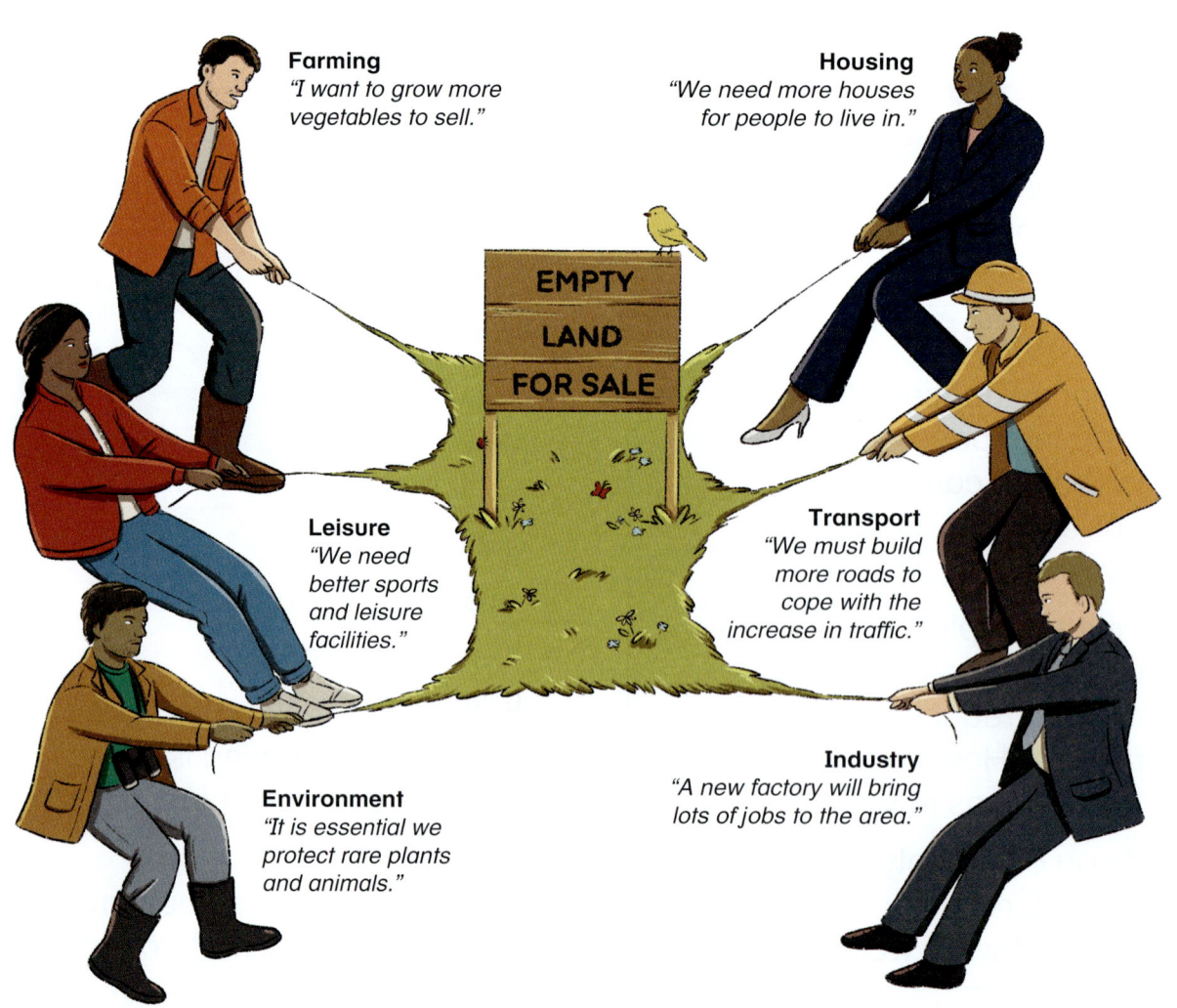

Farming
"I want to grow more vegetables to sell."

Housing
"We need more houses for people to live in."

Leisure
"We need better sports and leisure facilities."

Transport
"We must build more roads to cope with the increase in traffic."

Environment
"It is essential we protect rare plants and animals."

Industry
"A new factory will bring lots of jobs to the area."

Unit 4 Planning issues

Living on an island

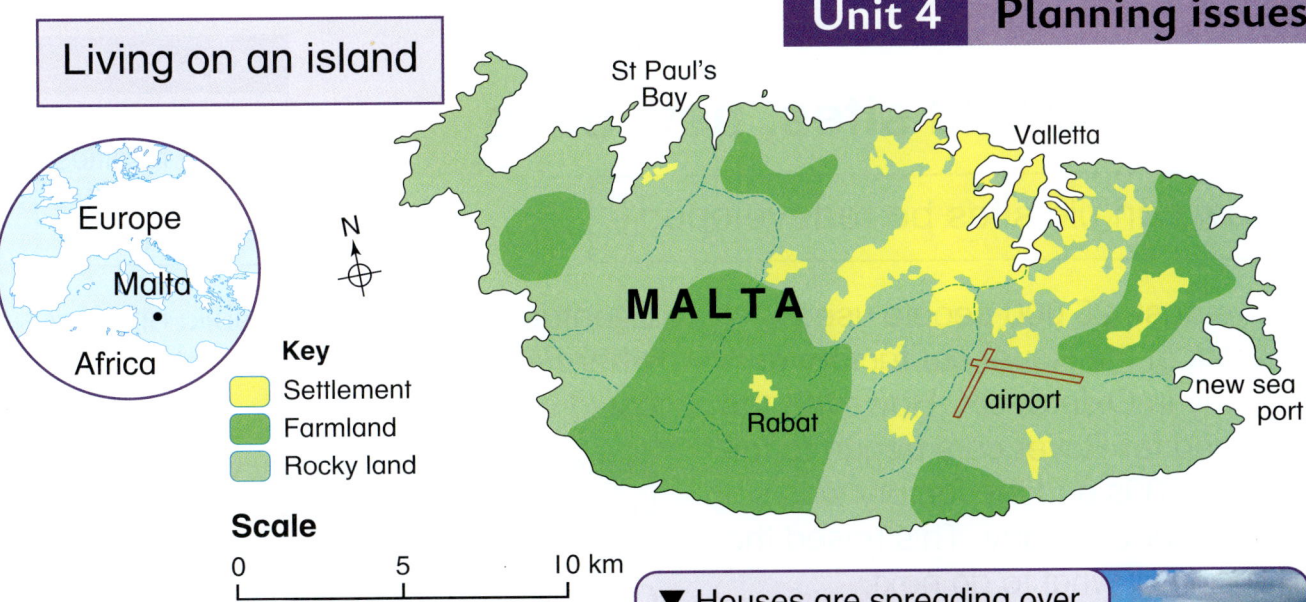

Key
- Settlement (yellow)
- Farmland (dark green)
- Rocky land (light green)

Scale: 0 — 5 — 10 km

Malta is a small island in Europe. It is in the Mediterranean Sea between Italy and North Africa. In the past, buildings covered only a small part of the island. Now they cover a third of the land. The new developments include:

- large, modern houses in country areas
- factories and warehouses
- a new port to encourage trade
- hotels for tourists
- greenhouses to grow vegetables
- rubbish dumps.

▼ Houses are spreading over the countryside in Malta.

The government has made strict planning laws to control development. New planning schemes have to be considered carefully.

Sustainability is a key issue. Having enough good water supplies is a major challenge. Disposing of waste is another problem. In the future people will have to make very careful decisions as there is little land to spare.

Discussion
- Look at the picture on page 20. Which type of land use do you think is most and least important?
- Talk about a site near you which could be redeveloped. What might it be used for?
- Why doesn't the government stop all new development in Malta?

Climate change
How do you think climate change could affect (a) tourism (b) farming in Malta?

Unit 4 Planning issues

Lesson 2: Old sites, new uses

How can old sites be redeveloped?

Key words
council
leisure facilities
public enquiry
redevelopment

In the past, 25 000 people used to work at the Rover car factory at Cowley in Oxford, UK. However, the demand for Rover cars began to fall in the 1980s and the factory was sold to another company. This raised the question of what to do next.

The discussions involved local people, the car company and the town council. There were regular reports in the local newspaper.

Eventually there was a public enquiry where everybody who was interested could give their opinion.

In the end the old factory was knocked down to make room for a new one which builds electric cars. Other areas were redeveloped as a business park for science and technology companies. There was also space for a hotel and supermarket.

▼ Workers on their way home from the Rover factory in the 1950s.

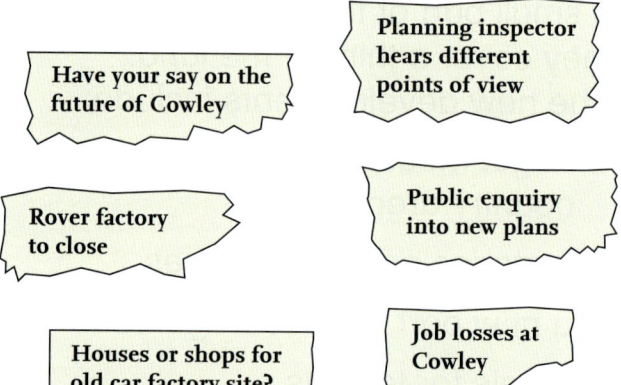

Have your say on the future of Cowley

Planning inspector hears different points of view

Rover factory to close

Public enquiry into new plans

Houses or shops for old car factory site?

Job losses at Cowley

Plan	Advantage	Disadvantage
Keep the existing factory open	Preserves jobs and keeps the factory in use	Old factory difficult to modernise and expensive to run
Close the factory and redevelop the land for housing	Helps to provide homes for the people of Oxford	Factory workers would lose their jobs
Use the land for new offices, shops, factories and a hotel	Creates over 4000 new jobs and provides shops for local people	Fails to create any new houses or leisure facilities

Discussion
- Why did the Rover factory close?
- What were the new plans?
- Do you think anything was missing from these plans?

Unit 4 Planning issues

▼ The factory builds Mini brand electric cars.

▼ Science and technology companies have been set up in the business park.

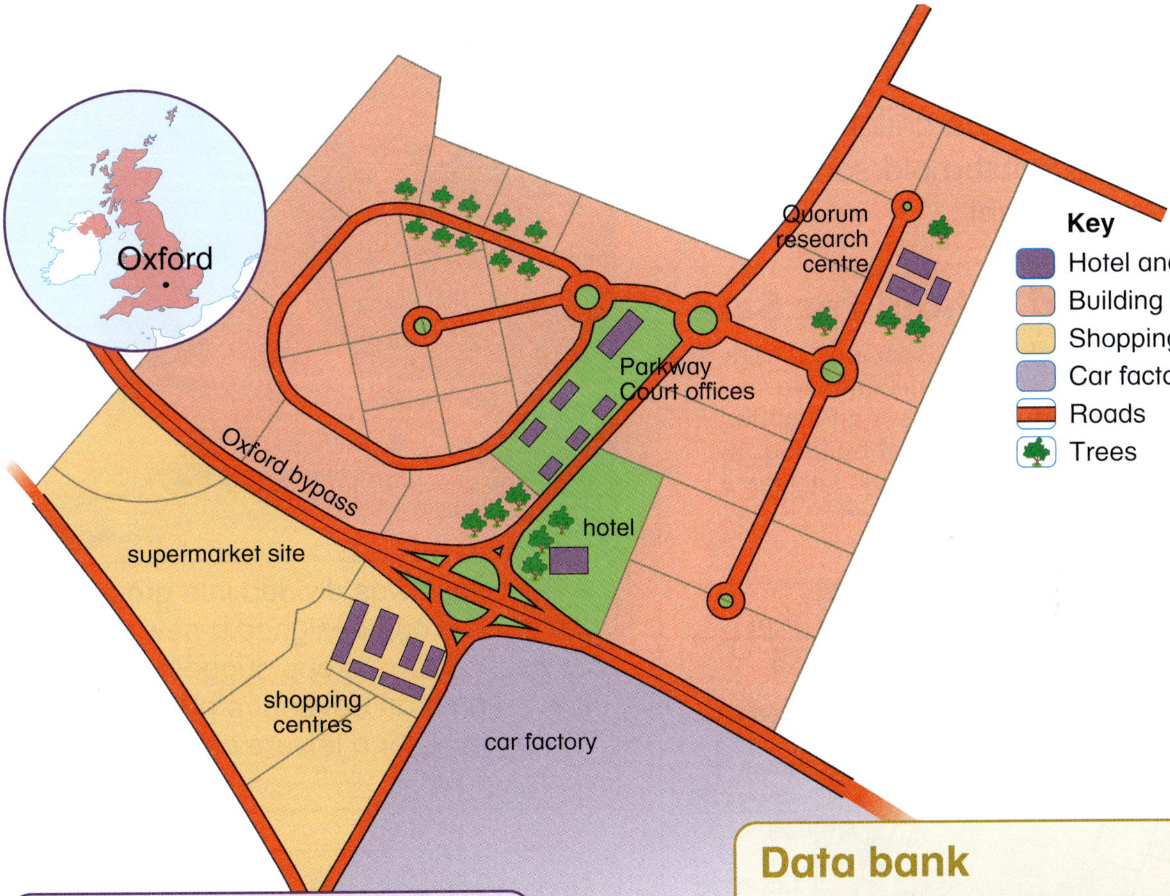

Key
- Hotel and offices
- Building land
- Shopping centre
- Car factory
- Roads
- Trees

Oxford

Quorum research centre
Parkway Court offices
Oxford bypass
supermarket site
hotel
shopping centres
car factory

▲ One of many plans for the site.

Investigation
Write sentences explaining (a) why the site needed to be redeveloped, (b) the different plans suggested, and (c) why the mixed development seemed best.

Data bank
- Town planning dates back to Roman times.
- Land which has been built on before is known as a brownfield site.
- Land very rarely goes back to being countryside once it has been developed.

Unit 4 Planning issues

Lesson 3: Planning game

How are planning decisions made?

Key words
aerial photograph
features
leisure centre
site

At Maryland Primary School, the children did a planning activity. They found out how the land around them was used by making a list of the things they could see on a map and an aerial photograph. Next their teacher asked them to think about what other things could be done with the school site. The children wrote an advert to sell the old school site for redevelopment.

Maryland School for Sale

Maryland School is very spacious. There are two big halls, 14 rooms and two gigantic playgrounds. This is an exciting chance to buy a very special building. It could be used for many things such as a health centre, hotel or museum.

▲ The class was divided into groups and each group designed a new plan for the site. One group suggested an ice rink, another an adventure playground. The scheme for a leisure centre was very popular.

▲ A redevelopment plan for the site.

Mapwork

Look at the map and photograph on page 25. Make a list of the numbered features.

Discussion

- What is shown on an aerial photograph that isn't on a map?
- What do you think are the best things about the Maryland School site?

Unit 4 Planning issues

▶ An aerial photograph of the area around Maryland Primary School.

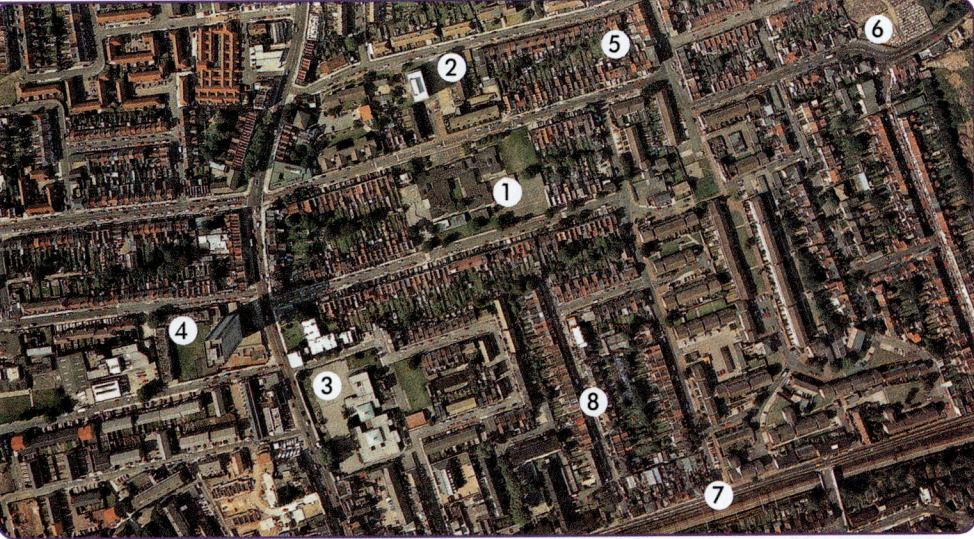

▶ A map of the same area.

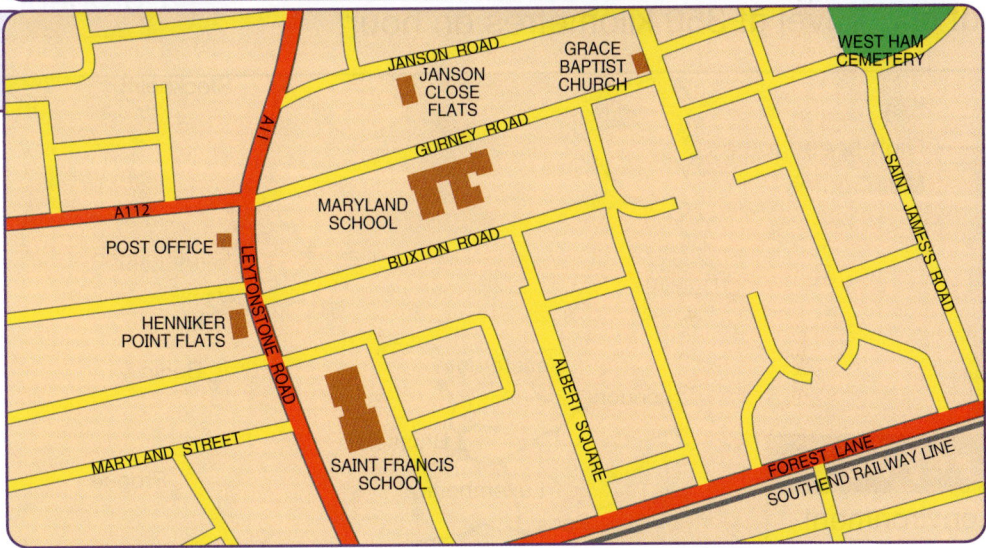

Investigation
Write a 'for sale' advertisement for the redevelopment of your old school site.

Mapwork
Imagine your school has moved and the old site is to be redeveloped. Devise a plan of your ideas for the site.

Summary
In this unit you have learnt:
- that people want to use land in different ways
- how planning decisions are made
- how to obtain information from maps and aerial photographs.

25

Unit 5 Transport

Lesson 1: Travelling further, travelling faster

What are the opportunities for travel in the world today?

The number of people in the world is increasing. At the same time, more and more people want to travel greater distances as easily as possible. Road, rail and air transport is constantly being improved to meet these needs. However, these changes are having a big impact on the environment.

Railways

Across Europe railways are being improved with a new high-speed network. Most of the main cities are now linked together with trains that can travel at 350 kilometres an hour.

Investigation

Devise a word search puzzle involving six European cities.

Key words

environment
network
transport

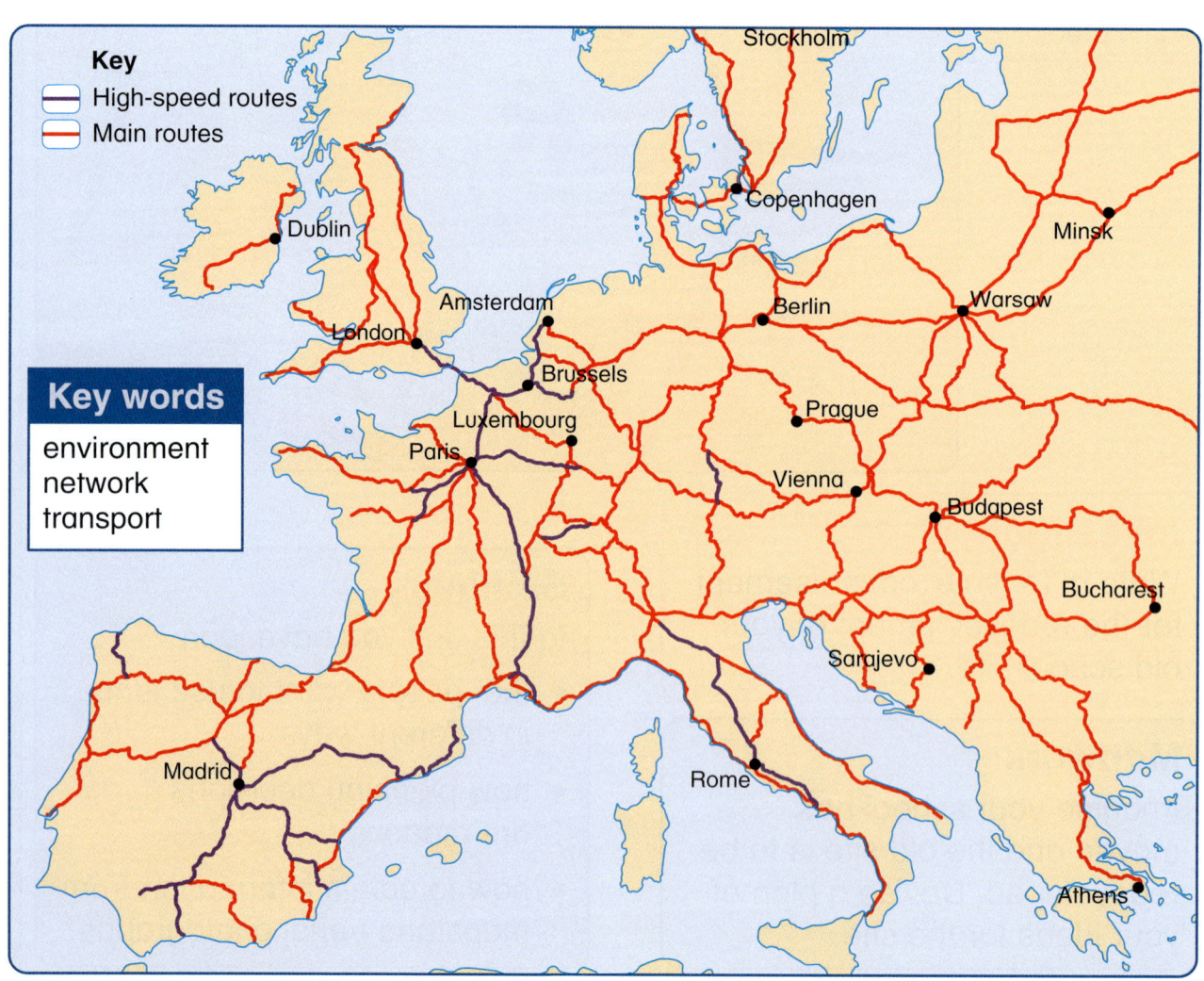

Unit 5 Transport

Air travel

Since they were invented a hundred years ago, aircraft have been getting steadily larger. The airbus can already carry over 500 people. In future even bigger planes and larger airports may be needed to cope with demand.

▲ Carbon emissions for a single person travelling in different ways from London to Glasgow in the UK.

Discussion
- Why do we need different types of transport?
- Which type of transport do you enjoy most?
- What is one good and one bad point about each different method?

Mapwork
Find and download an international air route map of your own. Write a sentence saying what it shows.

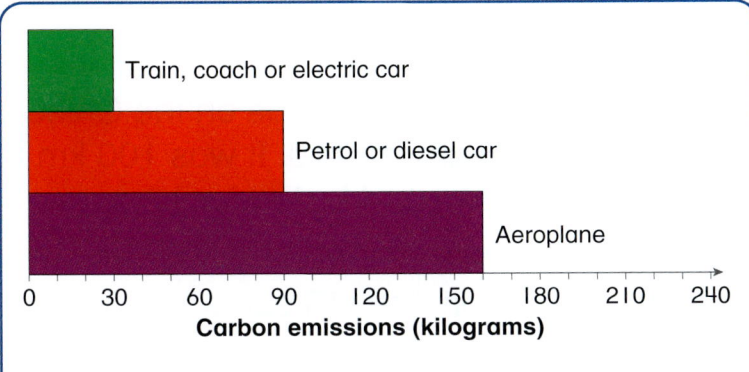

▲ Some of the air routes linking London to cities around the world.

Unit 5 Transport

Lesson 2: Transport problems

Can roads cope with more traffic?

Key words
bypass
flyover
traffic
vehicles

Cars are very convenient. They take us exactly where we want to go and can carry lots of luggage. Today there are more vehicles on the roads than ever before.

To help traffic move more easily, governments have spent huge amounts of money on improvements. However, keeping the traffic flowing is a huge challenge. Also, as roads are improved, they attract even more vehicles.

Data bank
- There are over 400 000 km of roads in the UK.
- One of the longest ever traffic jams was on the China Highway 110. It was 100 km long and lasted for 12 days.
- 28% of carbon emissions from the US come from transport.

Discussion
- What is convenient about cars?
- Why does the amount of traffic keep increasing?
- What do you think could be done about this?

Investigation
What problems for people and traffic are shown in the photograph? Make a list and suggest how you could solve each one.

▲ Sometimes large lorries have to go through the narrow streets of old, historic towns.

Unit 5 Transport

Traffic management

A lot of time and money has been spent trying to manage traffic problems. This diagram shows changes in the UK over the last hundred years.

British city streets blocked with traffic.

Roundabouts improve traffic flow.

Traffic lights control road junctions.

One-way streets introduced.

Underpasses and flyovers take traffic through urban areas.

Bypasses take traffic around villages and towns.

Motorway network built to link major cities.

Lower speed limits in villages and towns.

Cities set up 'low emissions zones' (ULEZ).

▼ London, UK, streets in the early 20th century.

▼ A traffic jam in central London in the 21st century.

Mapwork
Make a detailed plan of a local street. Add notes about the different rules which have been devised to keep people safe and traffic moving.

Unit 5 Transport

Lesson 3: Hidden costs

How do vehicles affect people and the environment?

Key words
action group
campaign
questionnaire
vehicle

All over the world, cars and lorries are damaging the environment.

Resources
Vehicles are expensive to make and use lots of resources. They are difficult to recycle when they wear out.

Air pollution
Petrol and diesel engines create exhaust fumes. All cars create dust from their tyres and brakes.

Health
Health problems such as asthma can be caused by air pollution.

Noise
Noise from cars and lorries cause stress for people who live near busy roads.

Wildlife
Motorways and bypasses cut through the countryside damaging plant and animal habitats.

Discussion
- What resources do cars use?
- In what way might Skye be damaged by the bridge?
- What is the worst problem caused by traffic?

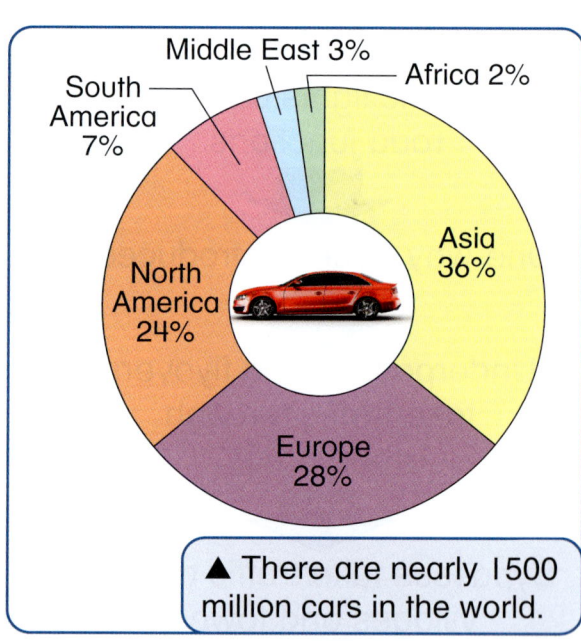

▲ There are nearly 1500 million cars in the world.

▼ Building a new bridge
Some people argue that the bridge which now links the island of Skye to mainland Scotland has spoilt its character.

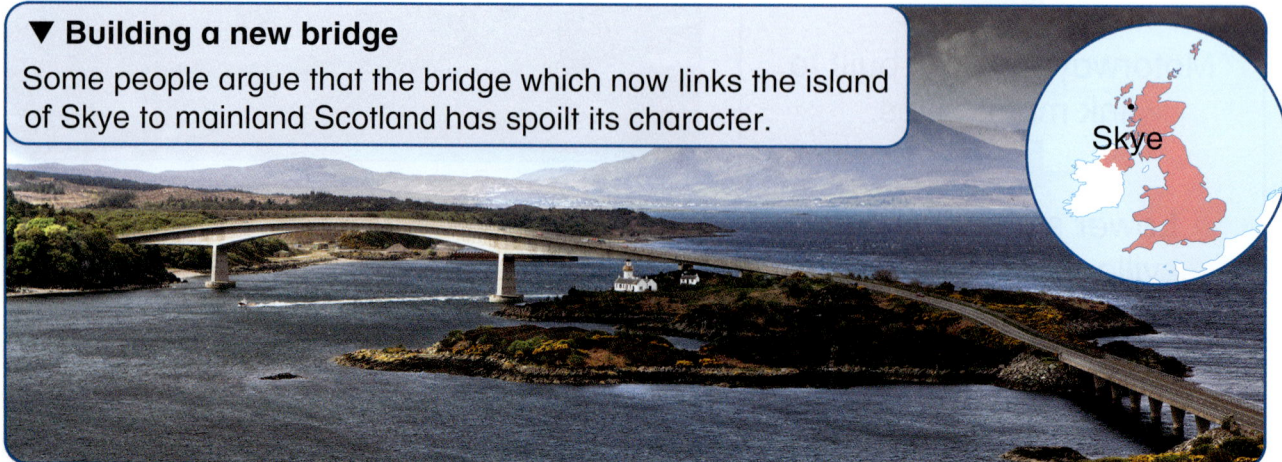

Unit 5   Transport

Finding out about local transport

At Vale View Primary School the children made a survey of traffic problems as part of a road safety campaign. First they made a list of ten problems. Then they devised a questionnaire to find out which was most serious. They asked passers-by in their local High Street for their views. The children coloured a box for each answer.

▲ Action groups monitor local traffic speeds

Traffic problems questionnaire

	1	2	3	4	5	6	7	8	9	10	11	12	13	14	15
Noise from traffic	■	■	■												
Not enough crossing places	■	■	■	■	■										
Traffic travelling too fast	■	■	■	■											
Roads with no pavements	■	■	■												
Not enough safety barriers	■	■	■	■	■										
Too many parked cars	■	■	■												
Shortage of cycle routes	■	■	■	■	■	■	■	■	■	■	■				
Exhaust fumes	■	■	■	■	■	■	■	■	■	■	■	■	■	■	
Heavy lorries	■	■	■	■	■	■									
Rush hour traffic jams	■	■	■	■	■	■	■	■	■	■	■	■	■		

▲ These are the results of the traffic project.

Investigation

Make a similar survey of traffic problems in the area around your school.

Summary

In this unit you have learnt:
- that traffic problems are difficult to solve
- about different schemes to control traffic
- how people can change their travel habits.

Unit 6 Conservation

Lesson 1: Threatened wildlife

Why are many plants and animals endangered?

Key words
endangered
extinction
mahogany
pesticides
teak

All over the world wildlife is threatened by people and the loss of habitat. Some animals are killed by accident because of pollution. Others are hunted for their skins. Plants and creatures also have less space as cities grow larger and forests are cleared for farming. Meanwhile, climate change is making it hard for some animals to survive.

Discussion
- Do you know of any plants and creatures that are endangered?
- How do you think climate change might threaten wildlife?
- Does it matter if a plant or animal becomes extinct?

◄ In some countries, tigers are worth more dead than alive because their bones are used to make medicines.

Unit 6   Conservation

There are probably about 30 million different plants and animals in the world today. Scientists fear that half the world's wildlife could disappear in the next 50 years. Tigers, elephants, whales and crocodiles are all endangered. So are many plants, trees and insects.

Different plants and animals are an essential part of the world in which we live. Many medicines are obtained from plants. We eat fruit and vegetables which once grew wild. However, large numbers of plants and animals could become extinct before we can learn anything about them.

Data bank
- Life first evolved on Earth around 3000 million years ago.
- The first fish evolved around 440 million years ago.
- Humans have lived on the Earth for the past 4 million years.

Investigation
Find out more about one threatened plant or creature. Write a short report and add pictures.

Mapwork
Draw small pictures of endangered animals such as tigers and turtles. Find out where they come from and pin them on a large world map as a class display.

▼ The World Wide Fund for Nature (WWF) and other groups are trying to save animals and plants from extinction.

AT RISK

Whales
Hunted for meat and oil.

 **Eagles**
Poisoned by pesticides.

Butterflies
Numbers declining as their habitat is destroyed.

Orchids
Dug up for house plants.

Rhinos
Killed for their horns.

 **Mahogany and teak**
Trees cut down to make furniture.

Unit 6 Conservation

Lesson 2: Antarctica

Why should Antarctica be conserved?

Key words	
iceberg	wilderness
satellite	world park
treaty	

Antarctica is the last great wilderness on Earth. It is covered by a huge sheet of ice and is cut off from the other continents by stormy oceans. There are over 50 research bases in Antarctica where scientists study the wildlife and the weather.

The layers of ice which have built up over thousands of years provide valuable clues about the climate long ago. Scientists have also discovered a remarkable hidden landscape under the ice which once had valleys, hills and forests.

▼ A satellite image of the continent of Antarctica.

Data bank
- Three-quarters of the world's fresh water is stored in the ice and snow of Antarctica.
- In winter the sea freezes around Antarctica for nearly 100 km.
- There are no trees on Antarctica.
- In 2000, the largest ever recorded iceberg broke off the Antarctic ice shelf (295 km long).

Discussion
- What makes Antarctica so special?
- Which seven countries claim parts of Antarctica?
- Do you think Antarctica should be a world park?

Unit 6   Conservation

Antarctica world park

There are valuable supplies of coal and iron ore under Antarctica. In addition, the surrounding seas are some of the best fishing grounds in the world. Some countries have tried to claim different parts of the continent for themselves. However, many people think that Antarctica is best left alone. They argue that mining will do serious damage to the environment and that fishing will make whales extinct.

In 1961, twelve countries approved a treaty agreeing:

- that Antarctica would only be used for peaceful purposes.
- that any claims to territory would be left undecided.

A total of 46 countries have now signed up to the treaty. However, the future of Antarctica still hangs in the balance. No one knows how long it will stay as a world park.

▼ Greenpeace campaigned for eight years to obtain a ban on mining in Antarctica.

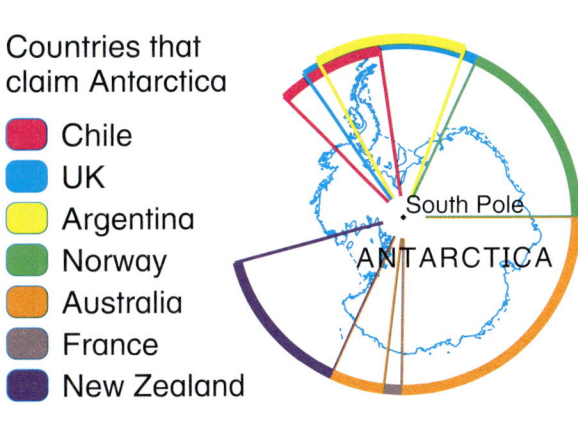

Countries that claim Antarctica
- Chile
- UK
- Argentina
- Norway
- Australia
- France
- New Zealand

Investigation

Find out about Captain Scott's journey to the South Pole. Write a page of notes with diary entries about what happened.

▼ A colony of emperor penguins with their chicks.

PROTECTED

Unit 6 Conservation

Lesson 3: Conservation projects

What are people doing to conserve the environment?

Every year, monarch butterflies spend the winter in the forests of Mexico. In spring they migrate north thousands of miles to the United States and Canada. However, butterfly numbers have been declining since the 1990s due to climate change and deforestation.

▲ Millions of butterflies in the Mexican forest.

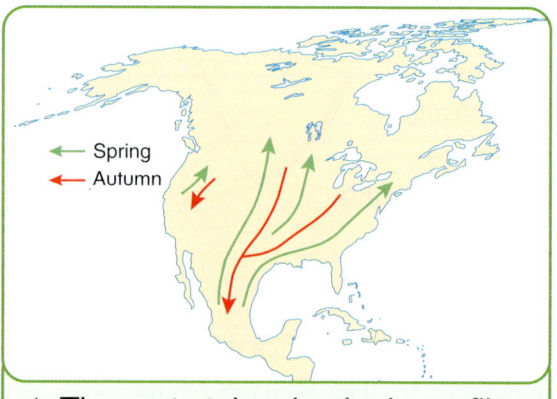

▲ The route taken by the butterflies

▼ Eco-tourists in the forest.

Eco-tourists come to see these amazing butterflies as they gather in their millions in the forest. This has created thousands of much-needed jobs. Local people no longer need to cut down the trees to earn a living and the forest is slowly being replaced. The monarch butterfly is no longer in danger.

Data bank
- The butterflies weigh about half a gram.
- They can fly over 150 km a day.
- The caterpillars only eat a plant called milkweed.

Discussion
- Why were monarch butterflies endangered?
- Why are butterfly numbers now increasing?
- Why was it important to involve local people?

Unit 6   Conservation

How can we keep a balanced environment?

Have you noticed that some fruit and vegetables in the supermarket have a label saying they are organic? This means that they have been grown without being sprayed by pesticides.

Key words

deforestation
eco-tourists
fertiliser
migrate
organic
pesticide

Jim and Pam Bennett run an organic farm in Aberdeenshire, Scotland. They grow oats, potatoes, carrots, swedes and grass. They also have 40 cows and some chickens.

No pesticides or harmful chemicals are used on the farm. Jim plants a different crop in each field every year which helps to keep the soil healthy. He also uses animal dung as fertiliser.

Organic farming is hard work. However people say the food tastes better. In addition, there are more deer, birds, badgers and butterflies in the fields.

"We have to think of the future of the world," Pam says. *"We want to care for the environment."*

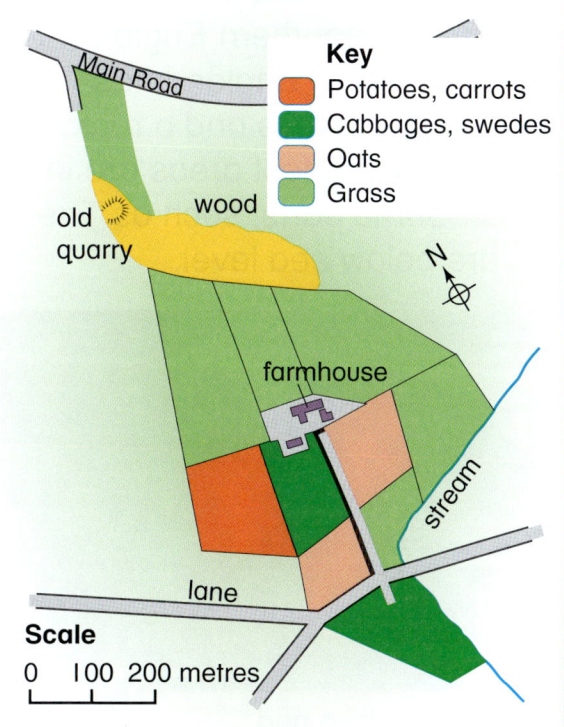

Mapwork
Make notes around a plan of your school or local area to show how it might be improved for wildlife.

Investigation
Find out about the advantages and disadvantages of organic farming. Draw illustrations to show them.

Summary
In this unit you have learnt:

- why wildlife is threatened
- how people and countries can co-operate to protect the environment
- how farmers can use the land without harming it.

Unit 7 England

Lesson 1: Introducing England

What is England like?

England is the largest of the four countries which make up the United Kingdom. There are mountains in the Pennines and Lake District. Low hills of chalk and limestone stretch across southern England. The south west of England has deep wooded valleys and a rocky coastline. The flattest areas are in the east. Some parts, such as The Fens, are below sea level.

Key words	
Birmingham	London
Dartmoor	Pennines
The Fens	River Thames
Lake District	

▼ Potatoes, flowers, fruit and vegetables all grow well in The Fens.

▲ Buttermere lake and mountains in the Lake District.

Discussion

- Which are the flattest parts of England?
- What do the maps tell you about each of the places listed in the 'key words' panel?
- Which area of England would you most like to visit?

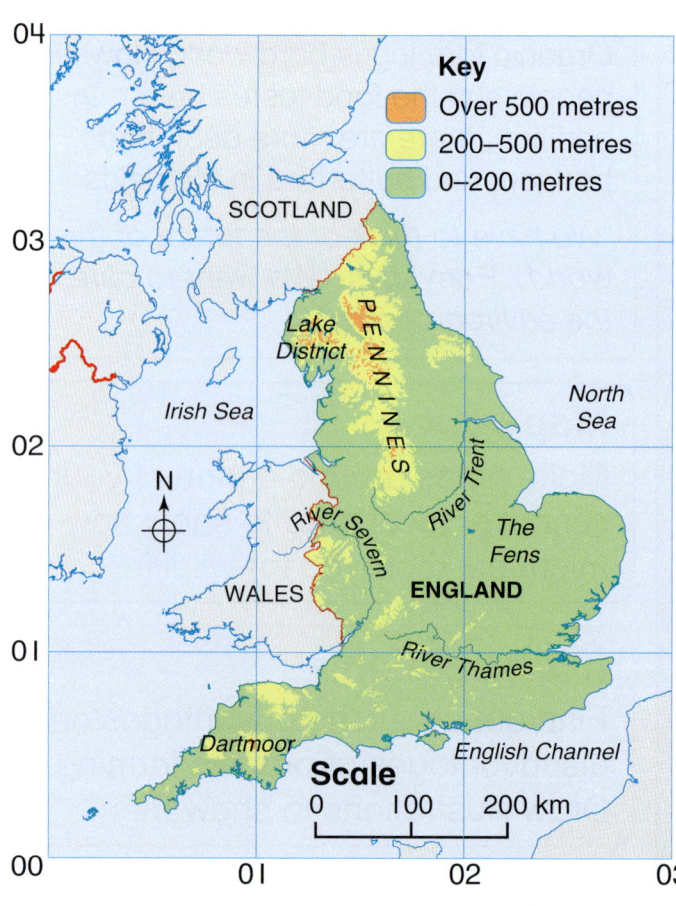

38

Unit 7 England

Blackpool Tower

Clifton Bridge, Bristol

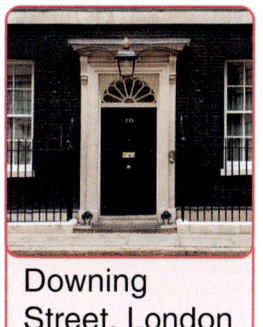
Stonehenge, Wiltshire

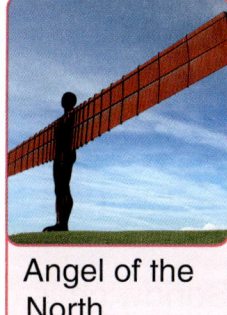
Downing Street, London

Angel of the North

Weather
The Lake District is the wettest area and eastern England is the driest.

The south coast has more sunshine than any other area in England.

Rivers and landscape
The Thames, Severn and Trent are the longest rivers in England.

The highest mountain is Scafell Pike in the Lake District (978 metres).

Transport
Motorways and railways spread out from London to all parts of the country. As many as 80 million people use Heathrow airport (24 km west of London) every year.

Settlement
More than three-quarters of the population live in towns and cities.

Southeast England is the most crowded area.

Work
London, the West Midlands and northern England are important industrial areas.

London is a worldwide centre for banking and insurance.

Investigation
Make a collage of ten pictures of England – two for each panel on this page.

Mapwork
Working from an atlas or online map, name ten English counties.

Unit 7 England

Lesson 2: Finding out about Sandwich

> How has Sandwich developed?

Key words	
bypass	port
industrial area	

Sandwich in history

13th century Sandwich was an important port, exporting wool to Europe.

15th century The River Stour silted up and the trade stopped.

20th century Barges loaded with explosives for the First World War sailed from a secret port just north of Sandwich.

Sandwich is a small town on the River Stour near the Kent coast. It was founded by the Anglo-Saxons over one thousand years ago who called it 'Sandwich' or 'village on the sands'.

Today Sandwich has a population of around 5000 people. The old streets and historic buildings attract tourists from the UK and abroad.

There have been many changes. Some years ago, there was a coal-fired power station and a large chemical factory on the edge of the town. This has now been redeveloped as a science park and industrial area. A bypass has been built round the edge of the town to keep traffic out of the centre. Meanwhile, otters, beavers and parakeets are now seen along the banks of the River Stour.

▼ Sandwich has many sheltered gardens.

▼ The River Stour flows along the edge of the town.

Discussion
- How long has there been a town at Sandwich?
- How has Sandwich changed?
- What do you think is special about Sandwich?

Unit 7 England

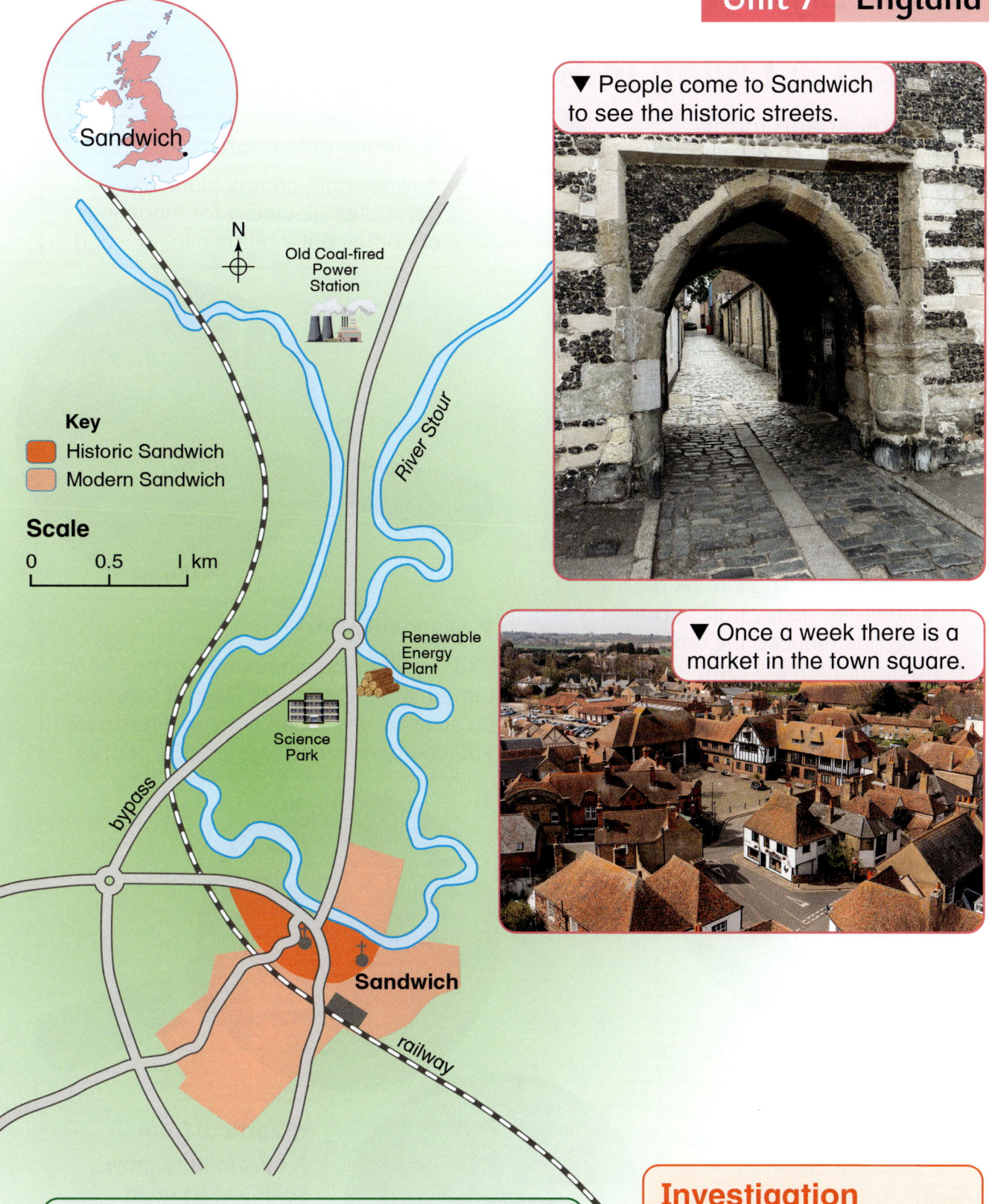

▼ People come to Sandwich to see the historic streets.

▼ Once a week there is a market in the town square.

Mapwork
Devise a short walk round your area for a friend or child from another school, stopping at six or more points of interest.

Investigation
Devise a timeline for your area showing major events and changes from long ago to today.

Unit 7 England

Lesson 3: Living in Sandwich

There have been many changes in and around Sandwich.

▲ **Traffic management**

The narrow, winding streets are unsuitable for heavy traffic. The bypass has reduced the amount of traffic through the town.

▼ **Renewable energy**

A renewable energy plant now generates electricity for thousands of homes but it burns a lot of wood.

▼ **Discovery Park**

The old chemical company site has been turned into a Discovery Park. It focuses on science and technology and has over 150 companies employing 3500 people.

▲ **New homes**

New homes have been built on old meadows along the banks of the River Stour. These include both apartments and houses.

◀ **Vertical farm**

A new factory grows salads stacked in layers under artificial light. This method saves water, cuts down on pesticides and reduces 'food miles'.

Unit 7 England

Quality of life

▶ People have different views about living in Sandwich.

Speech bubbles:
- I can buy all my weekly shopping in the town.
- There is nothing for young people to do in the evenings.
- I like the old buildings and quiet streets.
- I can walk to all the places I want to go to so I don't need a car.
- I get fed up with all the tourists in the summer.
- I am worried by all the fumes in the air.

Living in Sandwich

Shopping
Do the shops sell most of the things you need? ☐ Yes ☐ No

Transport
Is it easy to walk from place to place? ☐ Yes ☐ No

Leisure
Are there enough facilities for young people? ☐ Yes ☐ No
Are there enough facilities for older people? ☐ Yes ☐ No

Character
Do you find the town attractive? ☐ Yes ☐ No
Which features do you like best? ....

Environment
Is the town affected by any problems? ☐ Yes ☐ No
List the worst problems
....
....
....

Discussion

- What do the people in the picture like and dislike about living in Sandwich?
- Do you think all the changes are good for the town?
- Would you like to live in Sandwich?

Key words

| leisure | technology |
| renewable energy | food miles |

Summary

In this unit you have learnt:

- about the physical and human geography of England
- how photographs, maps and words can give you information about a place
- how to investigate the quality of life.

Investigation

Using questions from the 'Living in Sandwich' survey, find out what people think about living in your area.

Unit 8 Europe

Lesson 1: Introducing Europe

What are the regions of Europe?

Europe is about 4000 kilometres from north to south. It has a wide variety of landscapes. These include the vast forests of Russia and high mountain ranges, like the Alps.

Discussion
- What is the landscape for the three cities marked on the map?
- If you travelled across Europe from north to south, what differences would you see?
- How would you describe an area of Europe in just a few sentences?

Data bank
- Europe is the only continent without deserts.
- Europe has a remarkably long coastline – approximately 66 000 km.
- Finland is 75 per cent forested and has the most trees in Europe.

Key words
fjord
grassland
Mediterranean
tundra

44

Unit 8 Europe

Living in Europe

◀ Scandinavia

My name is Erika.

I live in Bergen in the west of Norway.

My father is the captain of a ferry boat. He takes the ferry along the coast stopping at villages on the fjords to deliver mail and other goods. People also travel on the ferries.

▶ Central Europe

My name is Hania.

I live in Prague, the capital city of Czechia.

My parents work in a factory which makes car parts. At weekends we often go out into the countryside to walk in the woods and hills.

◀ Mediterranean lands

My name is Carlos.

I live in Seville in southern Spain.

My mother works on a fruit farm. The long, hot summers and wet winters are good for growing oranges and lemons.

Mapwork

Using an atlas, find a country that contains three or more of the environments shown on the map on page 44.

Investigation

Find out about the countries you might pass through on a journey from Seville to Bergen.

Unit 8 Europe

Lesson 2: The European Union

How can countries work together?

Key words
agriculture
tax
trade

After the Second World War (1939–1945) much of Europe lay in ruins and had to be rebuilt. People decided that if they worked together it would help to keep the peace.

In 1957, France, Germany, Italy, Belgium, Luxembourg and the Netherlands agreed on a scheme to develop farming and industry and increase wages. This was the start of the European Union (EU).

To begin with there were just six countries in the EU. By 2025 there were 27 members with a total population of around 450 million people. The UK joined the EU in 1973 but left in 2020 after people voted in a referendum.

▼ Map of the countries in the European Union in 2024.

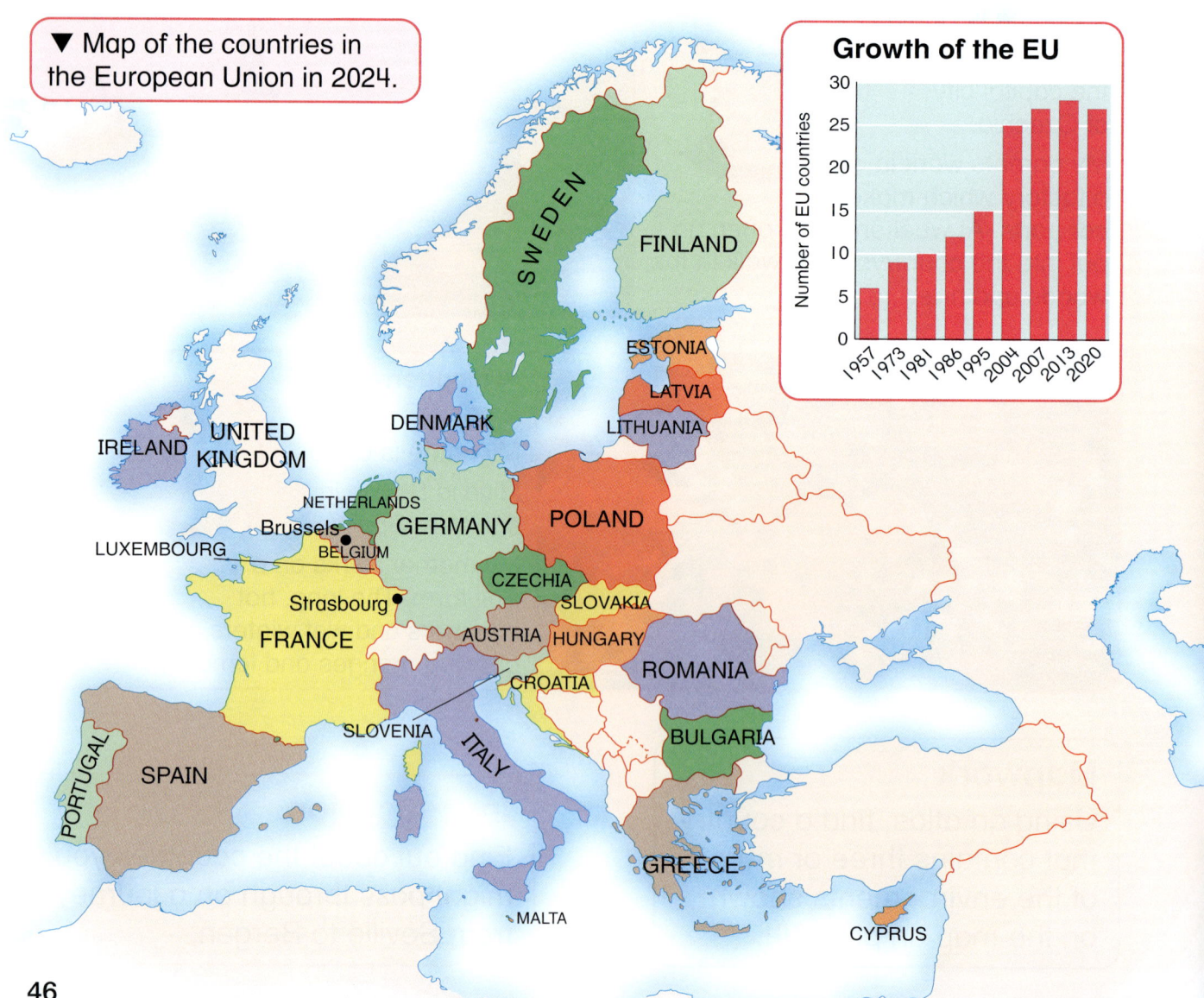

46

Unit 8　Europe

The European Union in action

The European Union works together to improve people's lives.

Trade　All the countries in the EU co-operate over trade. Producers do not have to pay a tax when they take their goods into another country.

Joint projects　Different countries in the EU work together on big, expensive projects. This includes building aeroplanes and scientific research.

Farming　The Common Agricultural Policy (CAP) tries to make sure that people receive food at reasonable prices and that farmers are properly paid for the goods they produce.

Working conditions　Within the EU, people can work in the country of their choice without having to have special permission. There are also laws about hours of work, safety and levels of pay.

Fishing　So many fish have been caught around Europe that stocks are running low. The EU has passed laws about when fishing boats can go out to sea, where to fish and the type of nets they can use.

Climate change　Many countries in the EU suffer from the same environmental problems. Laws have been passed to cut carbon emissions, reduce pollution and protect plants and creatures.

▲ These are some of the improvements that the European Union has achieved since it was formed over 50 years ago.

Discussion
- Why was the EU formed?
- How does the EU improve people's lives?
- Why might a country want to stay out of the EU?

Mapwork
Make an alphabetical list of countries in the EU.

Investigation
What do you think are the three best reasons for a country to join the EU? Write a sentence giving reasons for each choice.

Unit 8 Europe

Lesson 3: Celebrating Europe

What is special about Europe?

Last year, the children at St Mark's School organised a special European week. They searched online and looked in books to find out about different countries in Europe. People who had worked in Europe came into school to talk about their jobs and life abroad. Some of the children found out how to write sentences in foreign languages. Others looked at famous paintings and listened to folk music from other countries.

Data bank
- Europe is named after an ancient Greek princess, Europa.
- Vatican City in Rome is the smallest country in Europe.
- Europe and North America are moving a few centimetres apart each year.

- Bonjour — French
- God Morgen — Danish
- Guten Morgen — German
- ¡Buenos dias! — Spanish

▲ Different ways people say 'good morning' in Europe.

▼ More than 25 countries are shown in this satellite image of Europe. How many can you name?

Unit 8 Europe

All about Europe

▲ Discover how the island of Surtsey was created.

▼ Make a class scrapbook about some of the creatures that live in Europe.

▲ The Eiffel Tower in Paris is one of the most famous landmarks in the world. Try making your own tower model.

▼ Choose a country or city on the River Danube. How many words you can make from the letters in its name?

▼ Find out three facts about the Acropolis in Athens.

Investigation
Make your own 'Wonders of Europe' presentation, giving reasons for your choice.

Mapwork
Photocopy a map of Europe or make one of your own. Now cut it into around ten pieces. Challenge another child in your class to put it together again.

Summary
In this unit you have learnt:
- about the different countries and landscapes of Europe
- why the European Union was formed
- what makes Europe special.

Unit 9   South America

The Amazon

Lesson 1: Introducing the Amazon

What is the Amazon like?

The River Amazon is over 6400 kilometres long. It rises in the Andes and flows into the Atlantic Ocean. The Amazon has many tributaries flowing into it. The river basin covers a large part of South America.

The Amazon is on the equator where the climate is very wet and warm. Over many thousands of years a vast variety of plants and animals has developed. Around ten per cent of the world's living species are found in the Amazon rainforest. There are also indigenous communities living along the Amazon river. Today, they have to share the rainforest with settlers, who take advantage of the many natural resources found there.

Data bank
- Amazonia is nearly the same size as Europe.
- There are about 400 tribes in the Amazon, each with its own language, culture and land.
- About three-quarters of the food we eat originally came from the rainforests including rice, bananas, potatoes and tomatoes.

Key words

loggers	river basin
indigenous communities	species
	teak
rainforest	tributaries

▼ It has taken millions of years for the Amazon rainforest to evolve.

▲ There are 33 species of Amazon parrot.

Unit 9 — South America

Why is the rainforest being cleared?

◀ **Farming**
Soy bean farming now covers 8 million hectares of Amazonia. Soy is used as food for both humans and cattle. As land is used to grow soy beans, people are driven into the forest to clear new areas.

▶ **Cattle ranching**
More and more people want to eat meat. Amazonia has a good climate for cattle ranching. This has encouraged people to clear the rainforest to make grassland for animals.

◀ **Logging**
World demand for wood and paper has encouraged people to cut down trees. Sometimes loggers just remove valuable trees like teak but this also damages the forest.

▶ **Roads and mines**
Roads and highways are opening up more and more remote areas. In some places mines have polluted the land. In other places dams have flooded vast areas.

Mapwork
Using an atlas, make a list of countries around the Amazon river basin.

Investigation
Make up your own fact file about the Amazon.

Unit 9 South America

Lesson 2: Using the rainforest

What is it like to live in the rainforest?

Many communities live in the rainforest in Brazil. The forest provides them with food, the materials to build houses and many other things they need. Some people work as rubber tappers, collecting sap from rubber trees in the forest. They use the trees without damaging them.

Key words
- avocados
- latex
- manioc
- sap
- thatch

Discussion
- What is a rubber tapper?
- What resources do local people get from the rainforest?
- What might change the way of life for rainforest communities?

Some people in the Amazon live in wooden houses.

They grow rice, maize and manioc (which is like potatoes) in forest clearings.

When rubber tappers come to a rubber tree, they cut the bark and collect the sap, called latex in a cup.

Rubber tappers light fires and use the smoke to turn the latex they've collected into rubber.

There are snakes, monkeys and other animals in the forest.

People who live in the rainforest know how to use plants to make medicines.

People may catch birds or animals to eat.

There are plenty of coconuts, avocados and other fruit for food.

Palm trees provide thatch for the house roofs.

52

Unit 9 South America

Why is the rainforest so important?

As the rainforest is cleared, people are realising they are losing a unique habitat. Once it is lost it will be lost forever.

> "Today everyone wants to make money out of the Amazon, and we are scared. Scared by the burning that is taking place, by the destruction that is taking place, by the pollution. I speak as a person who has lived in the forest all his life."
>
> Leader of the Kayapo Indians, an indigenous community in the Amazon rainforest.

Investigation

Write a few sentences explaining (a) how the trees protect the soil (b) what happens when the forest is cleared.

Data bank
- Many plants and animals are becoming extinct before scientists have even studied them.
- As the forest is cleared, people are losing their homes.

Climate change

A huge quantity of water evaporates from the trees in the Amazon rainforest. If too many trees are cut down, it could cause changes to the climate around the world.

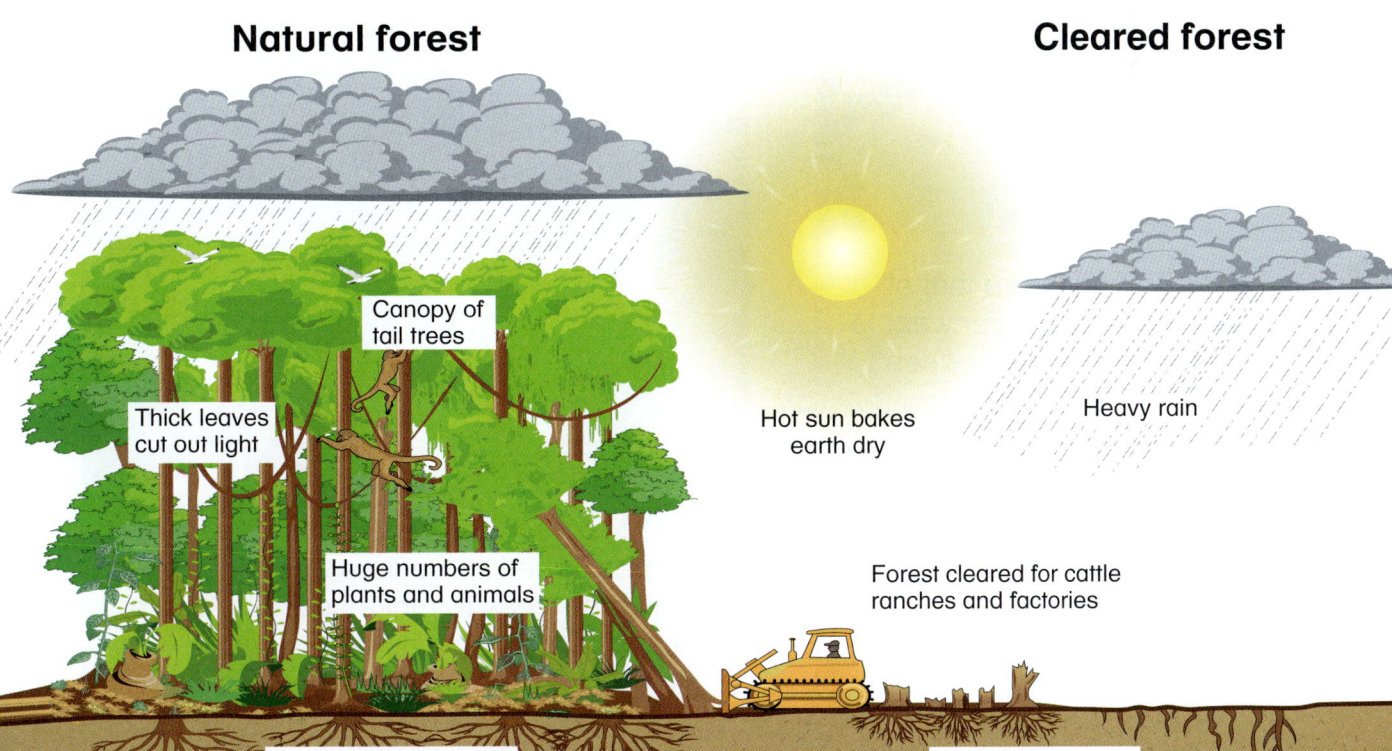

Natural forest — Canopy of tall trees; Thick leaves cut out light; Huge numbers of plants and animals; Roots protect the soil.

Cleared forest — Hot sun bakes earth dry; Heavy rain; Forest cleared for cattle ranches and factories; Soil washed away.

Unit 9   South America

Lesson 3: Saving the Amazon

What was Chico Mendes trying to do?

Key words	
cattle	reserves
plantation	rubber
ranch	tapper

Chico Mendes was brought up in Brazil collecting rubber from the trees. He is famous for trying to save the Amazon rainforest. He managed to persuade the government to set up reserves managed by local people. This was a new idea at the time. Here is his life story.

▲ ② When I was 18 I was taught to read and write by a friend. We used to listen to the radio to find out what was happening in the world. Every night we talked about what we had heard.

▲ ① My life began like that of all rubber tappers. I never went to school and I started work when I was nine years old.

▲ ③ I realised we needed to save the rainforest. The landowners were busy burning the trees to make space for cattle ranches. I joined the Rubber Workers Union. The leaders of the local American Indian groups agreed to help as well.

Discussion
- How did Chico Mendes learn about the world?
- What did he want the government to do?

Unit 9 South America

▲ ④ We asked the government to set up reserves where people could use the forest without damaging it.

▲ ⑥ After this I realised my life was in danger. However, I knew I had to go on trying to save the rainforest.

▲ ⑤ At about this time the plantation where I worked was sold to a new owner. His name was Da Silva. He tried to drive us off the land so he could set up a ranch but we stood firm.

Data bank
- In the rainy season the Amazon can be as much as 140 km wide.
- There are no bridges across the Amazon.
- Thirty-six million hectares of forest have now been protected.

Investigation
Draw a timeline showing Chico Mendes' life.

Mapwork
Working from an atlas or the internet draw a map showing rainforest areas around the world.

Summary
In this unit you have learnt about:
- why the Amazon rainforest is so special
- how it is threatened
- how it can be protected.

Unit 10 Asia

Lesson 1: Southeast Asia

What is Southeast Asia like?

Key words

colony	peninsula
natural resources	Tropics
palm oil	typhoons

Southeast Asia consists of a mixture of islands, peninsulas and coastal areas. It lies within the Tropics and was once heavily forested. Today great cities have sprung up across the region. This has made it one of the most influential regions of the modern world.

Indonesia is the largest country in Southeast Asia. It has many volcanoes and high mountains. Further north, the Philippines is another island nation. Here typhoons are sometimes a hazard.

In the past, many countries in Southeast Asia were European colonies. Since independence they have achieved huge amounts of development. Factories now produce cars, clothes and electrical goods. Palm oil is a key crop. Oil, tin and gold remain important natural resources. Development was so fast, the countries were nicknamed 'tiger economies'.

Discussion

- What are the landscape features of Southeast Asia?
- What vegetation would you expect to find in Southeast Asia?
- What is a 'tiger economy'?

Unit 10　Asia

▲ There are over 50 active volcanoes in Indonesia.

▶ As palm oil plantations replace the rainforest, many plants and creatures have lost their homes.

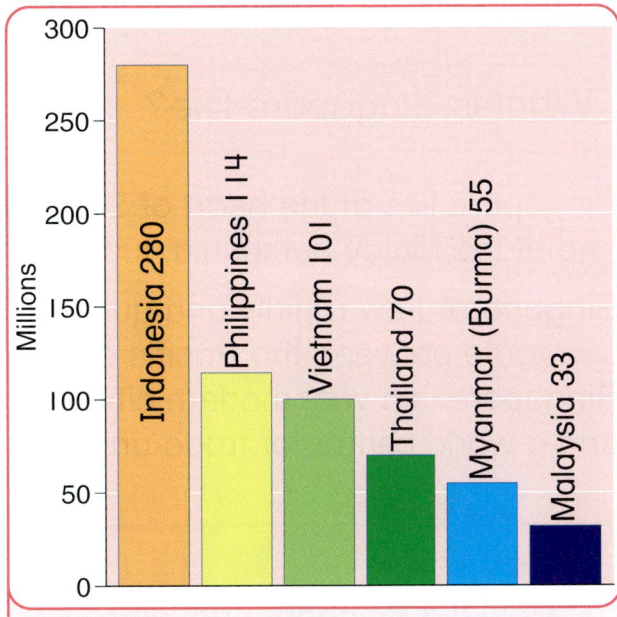

▲ Southeast Asia has roughly the same population as Europe.

Mapwork
Mark some of the largest cities on your own map of Southeast Asia.

Investigation
Devise a fact file for one Southeast Asian country giving details of the landscape, cities, products and environment.

▼ A night time view of Kuala Lumpur, Malaysia.

Unit 10 — Asia

Lesson 2: Investigating Singapore

What is Singapore like?

Singapore lies at the heart of Southeast Asia at the southern end of the Malay peninsula, just north of the equator.

Singapore grew rapidly and quickly came to dominate the sea route between the Pacific and Indian Oceans. Today Singapore is a vast modern city with over 6 million people and a world centre for trade and finance.

Discussion
- How did Singapore develop?
- What is Singapore like today?
- What makes Singapore special?

Nature Reserve
The rainforest which once covered Singapore is preserved in the Bukit Timah Nature Reserve.

New towns
There are 24 new towns in Singapore.

Bridge
Two busy causeways link Singapore to Malaysia, providing routes for people, traffic and water pipes.

Docks
Singapore docks handle 130 000 containers a year.

Industry
There are over 100 oil and chemical companies on Jurong Island.

Reclaimed land
As land is reclaimed Singapore island is growing larger and changing in shape.

Unit 10 Asia

▼ Offices and tall buildings tower above historic warehouses.

▲ The plants and glass domes in the Gardens by the Bay attract thousands of visitors.

▼ Climate

Singapore has a tropical climate with heavy rain throughout the year. Climate change is making it even warmer.

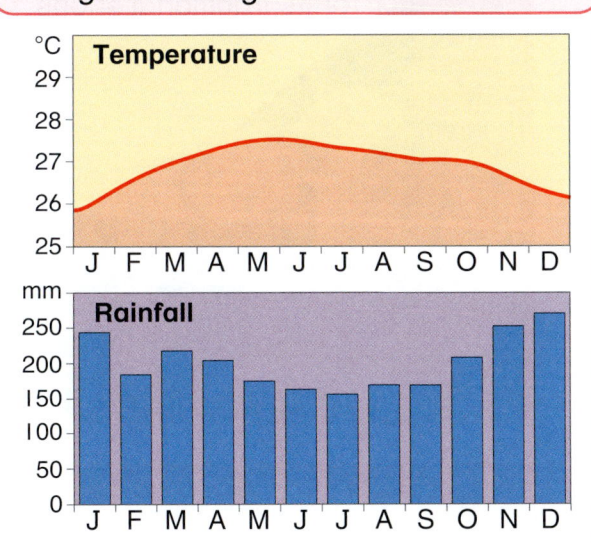

Data bank
- Singapore is the smallest country in Asia.
- Singapore is the most densely populated country in the world apart from Monaco.
- The last wild tiger was shot in Singapore less than a century ago.

Airport
With around 60 million passengers a year Changi Airport is one of the busiest in the world.

Investigation
Devise an advertisement encouraging a business to set up in Singapore.

City centre
Singapore is one of the world's largest financial centres.

Key words
causeway
equator
finance
peninsula
reclaimed land
tropical

Mapwork
Working from an atlas, name some other places which, like Singapore, are close to the equator.

Unit 10 Asia

Lesson 3: A Singapore family

> What is it like to live in Singapore?

The Chow family moved to Singapore from southern China around 80 years ago. They came to escape war and arrived with only a few possessions. However the family soon found a new life. Mr Chow set up a shop as a tailor in the Chinatown area of Singapore.

Mr Chow and his wife soon had children. One of Mr Chow's sons found work in the shipping industry. He was promoted to take charge of work building the rigs and platforms which are used at sea.

Discussion
- What job did each member of the Chow family do?
- How do these jobs link with the history of Singapore?
- What do you think is the most important issue for Singapore in the future?

▼ Chinatown is famous for its traditional shops.

▼ Ships carry cargoes to and from Singapore.

Key words
| community | desalination plant |
| mangrove swamp | self-sufficient |

Mr Chow has now died but his grandchildren and relatives still live in Singapore. Many of them have been to university. One of his great nieces, Jessica, works for the Singapore port authority planning new developments.

▶ Jessica Chow with her family at her graduation from university.

Unit 10 Asia

Planning for the future

There is very little spare land in Singapore, so providing food, water and housing is a big challenge. The government is making careful plans for the future.

Tampines new town

Tampines is built on land which was once forest and mangrove swamp. There are flats and shops in the centre and factories and offices round the edge. Frequent automatic trains link Tampines with other parts of Singapore. There are plans for more self-sufficient towns like Tampines, each housing 200 000 people.

Drinking water

There are plans to make Singapore self-sufficient in water. A dam has been built across the bay to catch the rain that falls on the city centre and surrounding areas. Desalination plants have also been built to remove salt from sea water.

▼ The Marina Bay Barrage

▼ Tampines has its own shops, schools and parks.

Mapwork

Devise a plan of how you could arrange six blocks of flats around a precinct. Remember to include paths, gardens, shops and meeting places.

Summary

In this unit you have learnt:
- how Southeast Asia is changing
- about different aspects of Singapore
- how Singapore is planning for the future.

Climate change

The government has drawn up plans to deal with extreme weather events such as flash floods which can happen very quickly.

Glossary

Carbon emissions
Pollution created by burning fossil fuels, causing global warming.

Clay
A waterproof soil made of very small particles.

Climate
The pattern of weather over many years.

Colony
A country that has been taken over and exploited by another country.

Council
A group of people (often elected) who make decisions for a community.

Extinction
Species of plants and animals become extinct when they have all died out.

Flint
Layers of very hard, knobbly stones which build up in chalk and limestone.

Food miles
The distance food travels from the farm where it is produced to the shop where you buy it.

Fossil fuels
Fuels like oil and gas that create carbon emissions.

Glacier
A thick sheet of ice which flows very slowly down slopes towards the sea.

Global warming
The long term warming of the Earth's surface caused by activities such as burning fossil fuels.

Granite
A very hard, pink or grey, volcanic rock which is used in buildings and roads.

Greenhouse
A building with lots of glass where plants can benefit from light and warmth.

Greenhouse effect
The way that carbon dioxide and other gasses stop heat from the sun bouncing back into space from the Earth.

Indigenous communities
The people who are descended from the first inhabitants of an area.

Limestone
A rock which forms in shallow seas from the remains of countless sea creatures.

Net zero
Seeing that any activity which emits carbon dioxide is balanced by an activity that takes the same amount out.

Peninsula
A narrow strip of land which extends into the sea.

Public enquiry
A meeting where people can discuss a problem in front of an inspector.

Regenerative farming
Farming in a way that cares for the soil and works in harmony with the seasons and nature.

River basin
The area which is drained by a river and its tributaries.

Teak
A hard and very valuable wood that comes from the rainforest.

Treaty
An agreement between different countries.

Tropics
Parts of the world where the sun is directly overhead at least once a year.

Typhoon
A very violent tropical storm which brings gales and flooding.

Volcano
An opening in the Earth's crust where red hot, underground rocks and gas break to the surface.

Index

Acropolis, The, Greece 49
airports 27, 39, 59
Alps 44
Amazon rainforest 50–55
Amazon river 50
Antarctica 8, 34–35
atmosphere 18

Bolivia 11
building materials 7
bypasses 29, 30, 40, 42

carbon emissions 15, 18, 19, 27, 28, 47
carbon footprint 19
cars 15, 19, 22, 23, 27, 28, 30, 56
cattle ranching 51, 54
Chico Mendes 54–55
climate 34, 50, 51, 59
climate change 4, 7, 10, 12, 14–19, 21, 32, 36, 47, 53, 59, 61
coal 6, 15, 18, 19, 35, 40
colony 56
community garden 19
conservation 32–37
Cotswolds, UK 6
Croatia, caves 5
crops 12, 16, 20, 37, 56
Czechia 45

dams 11, 12, 51, 61
Danube, river 49
deserts 5, 6

earthquakes 2, 3
Earth's crust 2, 3
Eiffel Tower, France 49
endangered wildlife 32, 33, 36
England 17, 38–43
erosion 4
Europe 9, 21, 26, 30, 40, 44–49
European Union 46–47
Eryri (Snowdonia), UK 6

factories 8, 15, 18, 20, 21, 22, 23, 40, 42, 56, 61
farming 18, 20, 32, 37, 42, 46, 47, 51
Fens, The, UK 38
food miles 42

fossil 2
frost 4
fumes 30, 31, 43

glaciers 4, 5
global warming 14, 15, 16, 18
Greenpeace 35
greenhouse effect 15

habitats 30, 32, 33, 36, 53
health 30
Herefordshire, UK 6

India 11
Indonesia 56, 57
Indus, river 12
Isle of Skye, UK 30

jobs 20, 22, 36, 42, 48, 60

Kuala Lumpur, Malaysia 57

Lake District, UK 38, 39
land use 20–25
landscapes 4, 6, 34, 39, 44, 45, 49, 56, 57
limestone 5, 6, 7, 38
logging 51
lorries 28, 30

Malta 21
maps 10, 14, 21, 23, 24, 25, 26, 27, 36, 37, 38, 39, 41, 44, 46, 56, 58
Matterhorn, Switzerland 5
monarch butterflies 36
Monument Valley, US 5
motorways 28, 29, 30, 39
mountains 2, 3, 4, 6, 9, 12, 38, 39, 44, 56

Netherlands 46
net zero 18, 19
Niagara Falls, 5
noise 30, 31
Norway 45

oil rig 6
Oxford, UK 22, 23
organic farming 37

Pacific Ocean 2, 3, 58
pesticides 33, 37, 42
Philippines 56, 57
planning schemes 21, 22–25
plants 11, 12, 13, 14, 15, 16, 19, 20, 30, 32, 33, 36, 50, 52, 53, 59
pollution 12, 18, 19, 30, 31, 32, 47, 53

rain 4, 8, 9, 11, 12, 13, 16, 17, 53, 55, 59, 61
regenerative farming 18
reservoirs 9, 17
resources 8, 30, 50, 56
rivers 4, 5, 6, 8, 9, 10, 12, 39, 40, 42, 49, 50
rocks 2, 3, 4, 5, 6–7

Sandwich, UK 40–43
saving water 12, 13
Scafell Pike, UK 39
Scotland, UK 9, 30, 37
seismograph 2
Severn, river 39
Singapore 58–61
Southeast Asia 56–61
Spain 45
Stour, river 40, 41, 42
Switzerland 5

temperature 14, 15, 17, 59
Thames, river 8, 38, 39
trade 21, 40, 46, 47, 58
traffic 20, 28, 29, 30, 31, 40, 42, 58
transport 26–31
typhoons 56

United Kingdom 38

volcanoes 2, 3, 6, 56, 57

water 8–13
water supplies 10, 11, 21
waterworks 8, 9
waves 4
weather 17
wells 9, 10, 11
wind 4, 5
World Wide Fund for Nature (WWF) 33

63

William Collins' dream of knowledge for all began with the publication of his first book in 1819.

A self-educated mill worker, he not only enriched millions of lives, but also founded a flourishing publishing house. Today, staying true to this spirit, Collins books are packed with inspiration, innovation and practical expertise.
They place you at the centre of a world of possibility and give you exactly what you need to explore it.

Published by Collins
An imprint of HarperCollins*Publishers*
The News Building, 1 London Bridge Street, London,
SE1 9GF, UK

HarperCollins*Publishers*
Macken House, 39/40 Mayor Street Upper, Dublin 1,
D01 C9W8, Ireland

Browse the complete Collins catalogue at
collins.co.uk

© HarperCollinsPublishers Limited 2025

Maps © Collins Bartholomew 2025

10 9 8 7 6 5 4 3 2

ISBN 978-0-00-872833-5

All rights reserved. No part of this publication may be reproduced, stored in a retrieval system, or transmitted in any form by any means, electronic, mechanical, photocopying, recording or otherwise, without the prior written permission of the Publisher or a licence permitting restricted copying in the United Kingdom issued by the Copyright Licensing Agency Ltd, 5th Floor, Shackleton House, 4 Battle Bridge Lane, London SE1 2HX.

Without limiting the exclusive rights of any author, contributor or the publisher, any unauthorised use of this publication to train generative artificial intelligence (AI) technologies is expressly prohibited. HarperCollins also exercise their rights under Article 4(3) of the Digital Single Market Directive 2019/790 and expressly reserve this publication from the text and data mining exception.

British Library Cataloguing-in-Publication Data

A catalogue record for this publication is available from the British Library.

Authors: Stephen Scoffham and Colin Bridge (with additional original input from by Terry Jewson)
Publisher: Laura White
Product manager: Natasha Paul
Development editor: Judith Walters
Copyeditor and proofreader: Catherine Dakin
Cover designer and illustrator: Steve Evans
Internal illustrators: Jouve India Private Ltd and
Hannah Drennan, Beehive Illustration
Typesetter: David Jimenez
Production controller: Alhady Ali
Printed and bound in Great Britain by Bell and Bain Ltd, Glasgow

This book is produced from independently certified FSC™ paper to ensure responsible forest management.

For more information visit: www.harpercollins.co.uk/green collins.co.uk/sustainability

Acknowledgements

The publishers gratefully acknowledge the permission granted to reproduce the copyright material in this book. Every effort has been made to trace copyright holders and to obtain their permission for the use of copyright material. The publishers will gladly receive any information enabling them to rectify any error or omission at the first opportunity.

P6cbl riccardo Mancioli/Alamy Stock Photo; P10 map: IHME, Global Burden of Disease (2024) – with minor processing by Our World in Data; P11tr © cstepin; P17tr, P31, P40tr, P40br, P41tr, P42tr, P42cr, P42b, P60br © Stephen Scoffham; P22 © British Motor Industry Heritage Trust; P28b © Mick Lobb/Geograph; P29t The Print Collector/Alamy Stock Photo; P49tl ARCTIC IMAGES/Alamy Stock Photo; P54tl © Miranda Smith Productions Inc./Wikimedia.

All other photos Shutterstock.